Algebra, Topology, and Category Theory

A Collection of Papers in Honor of Samuel Eilenberg

Algebra, Topology, and Category Theory

A Collection of Papers in Honor of Samuel Eilenberg

Edited by ALEX HELLER
DEPARTMENT OF MATHEMATICS
THE GRADUATE SCHOOL
CITY UNIVERSITY OF NEW YORK
NEW YORK, NEW YORK

MYLES TIERNEY
DEPARTMENT OF MATHEMATICS
RUTGERS UNIVERSITY
NEW BRUNSWICK, NEW JERSEY

ACADEMIC PRESS New York San Francisco London 1976
A Subsidiary of Harcourt Brace Jovanovich, Publishers

ACADEMIC PRESS, INC.
111 Fifth Avenue, New York, New York 10003

United Kingdom Edition published by
ACADEMIC PRESS, INC. (LONDON) LTD.
24/28 Oval Road, London NW1

Library of Congress Cataloging in Publication Data

Main entry under title:

Algebra, topology, and category theory.

"Many of these papers . . . were read at a conference held May 3 and 4, 1974, on the occasion of Professor Eilenberg's sixtieth birthday, at the Graduate Center of the City University of New York."

"Published works of Samuel Eilenberg": p.

CONTENTS: Auslander, M. Large modules over artin algebras.–Chen, K.-T. Reduced bar constructions on deRham complexes.–Cheng, C. C., and Mitchell, B. Flatness and projectivity of modules that come from ℂ-sets. [etc.]

1. Algebra–Addresses, essays, lectures. 2. Topology–Addresses, essays, lectures. 3. Categories (Mathematics)–Addresses, essays, lectures. 4. Eilenberg, Samuel–Bibliography. I. Heller, Alex, (date) II. Tierney, Myles. III. Eilenberg, Samuel.
QA155.A53 512 75-30467
ISBN 0–12–339050–8

PRINTED IN THE UNITED STATES OF AMERICA

Contents

List of Contributors

Numbers in parentheses indicate the pages on which the authors' contributions begin.

MAURICE AUSLANDER (1), Department of Mathematics, Brandeis University, Waltham, Massachusetts

KUO-TSAI CHEN (19), Department of Mathematics, University of Illinois at Urbana-Champaign, Urbana, Illinois

CHARLES CHING-AN CHENG (33), Department of Mathematics, Oakland University, Rochester, Michigan

ELDON DYER (45), Department of Mathematics, The Graduate School, City University of New York, New York, New York

PETER FREYD (55), Department of Mathematics, University of Pennsylvania, Philadelphia, Pennsylvania

JOHN W. GRAY (63), Department of Mathematics, University of Illinois at Urbana-Champaign, Urbana, Illinois

DALE HUSEMOLLER (77), Department of Mathematics, Haverford College, Haverford, Pennsylvania

D. M. KAN (95), Department of Mathematics, Massachusetts Institute of Technology, Cambridge, Massachusetts

F. WILLIAM LAWVERE (101), Department of Mathematics, State University of New York at Buffalo, Amherst, New York

SAUNDERS MAC LANE (133), Department of Mathematics, University of Chicago, Chicago, Illinois

BARRY MITCHELL (33), Department of Mathematics, Rutgers University, New Brunswick, New Jersey

JOHN C. MOORE (145), Department of Mathematics, Princeton University, Princeton, New Jersey

JOHN RHODES (149), Department of Mathematics, University of California, Berkeley, California

GEORGE S. RINEHART† (169), Department of Mathematics, Cornell University, Ithaca, New York

† Deceased.

ALEX ROSENBERG (169, 181), Department of Mathematics, Cornell University, Ithaca, New York

J. T. STAFFORD (181), School of Mathematics, The University of Leeds, Leeds, England

MYLES TIERNEY (189, 211), Department of Mathematics, Rutgers University, New Brunswick, New Jersey

BRET TILSON (149), Department of Mathematics, City University of New York, Queens College, Flushing, New York

A. T. VASQUEZ (45), Department of Mathematics, The Graduate School, City University of New York, New York, New York

Preface

The editors take great pleasure in offering this collection of papers in honor of Professor Samuel Eilenberg. The variety of fields represented, algebraic topology, homological algebra, category theory, and automaton theory, will, we hope, begin to reflect the contributions and breadth of interest of a man whose work has profoundly influenced the course of modern mathematics.

Many of these papers, which are all by former students and associates, were read at a conference held May 3 and 4, 1974, on the occasion of Professor Eilenberg's sixtieth birthday, at the Graduate Center of the City University of New York. We thank all the participants of the conference, as well as the National Science Foundation for its generous support.

Large Modules over Artin Algebras

MAURICE AUSLANDER

Introduction

We recall that an artin ring Λ is said to be an artin algebra if its center C is an artin ring and Λ is a finitely generated module over C. Our main objective in this chapter is to point out various ways the representation theories of finitely generated and large (not finitely generated) modules over an artin algebra are related.

This work grew out of attempts to establish the converse of the following result (see [2] and [6]). Let Λ be a left artin ring such that there are only a finite number of nonisomorphic finitely generated indecomposable left Λ-modules, then every left Λ-module is a sum (direct, of course) of finitely generated indecomposable Λ-modules. While it is still an open question whether the converse is true for arbitrary left artin rings, the converse is settled here for artin algebras. In fact we establish the following stronger result:

Theorem A. For an artin algebra Λ the following statements are equivalent:

(a) There is only a finite number of nonisomorphic finitely generated indecomposable left Λ-modules.

(b) Every left Λ-module is a sum (direct) of finitely generated indecomposable Λ-modules.

(c) Every nonzero left Λ-module has a summand that is a finitely generated indecomposable Λ-module.

(d) If $M_0 \xrightarrow{f_0} M_1 \xrightarrow{f_1} M_2 \longrightarrow \cdots \longrightarrow M_i \xrightarrow{f_{i+1}} M_{i+1} \longrightarrow \cdots$ is a sequence of monomorphisms between finitely generated indecomposable left Λ-modules, then there is an integer n such that f_i is an isomorphism for all $i \geq n$.

(e) If $\cdots M_i \xrightarrow{f_i} M_{i-1} \longrightarrow \cdots \longrightarrow M_1 \xrightarrow{f_1} M_0$ is a sequence of epimorphisms between finitely generated indecomposable left Λ-modules, then there is an integer n such that f_i is an isomorphism for all $i \geq n$.

(f) Every indecomposable left Λ-module is finitely generated.

(g) Any one of the statements (a) through (f) with "left module" replaced by "right module."

The other main result of this paper is the following "local" version of Theorem A.

Theorem B. For a finitely generated indecomposable left module M over an artin algebra Λ, the following statements are equivalent:

(a) There is an infinite number of nonisomorphic finitely generated indecomposable left Λ-modules N such that $\operatorname{Hom}_\Lambda(M, N) \neq 0$.

(b) There is a denumerably generated large indecomposable left Λ-module N such that $\operatorname{Hom}_\Lambda(M, N) \neq 0$.

(c) There is a Λ-module N with no finitely generated indecomposable summands and such that $\operatorname{Hom}_\Lambda(M, N) \neq 0$.

It is particularly appropriate that these results appear here since the work depends in an essential way on homological and categorical techniques and ideas that indelibly bear Sammy Eilenberg's mark.

We would like to take this opportunity to thank L. Gruson for many helpful conversations.

1. Preliminary Results

By and large we follow the notation of [1], [2], [3], and [4], to which the reader is also referred for background on some of the techniques and ideas used here. For sake of completeness we devote much of this section to a summary of the pertinent parts of these papers. While some of the references given will be to statements about contravariant functors, these can be readily transformed to the desired statements about covariant functors by the usual duality arguments, a task left to the reader.

We assume throughout this section that Λ is a left artin ring. We denote the category of left Λ-modules by Mod Λ and the category of finitely generated left Λ-modules by mod Λ. Since mod Λ is a skeletally small category (i.e., the isomorphism classes of objects of mod Λ form a set), we can talk of the category of all additive functors from mod Λ to Ab, the category of abelian groups, which we denote by (mod Λ, Ab). For each M in mod Λ we denote the representable functory $X \to \mathrm{Hom}_\Lambda(M, X)$ for all X in mod Λ by $(M,\ )$. It is well known that $(M,\ )$ is a projective object in (mod Λ, Ab) for each M in mod Λ and that for each F in (mod Λ, Ab) there is a projective presentation that consists of an exact sequence in (mod Λ, Ab)

$$\coprod_{i\in I}(N_i,\ )\to\coprod_{j\in J}(M_j,\ )\to F\to 0$$

where the N_i and M_j are in mod Λ and $\amalg$ denotes (direct) sum. We recall (see [1]) that F is said to be finitely generated if the set J can be chosen to be finite, and F is said to be finitely presented if both the sets I and J can be chosen to be finite. In other words F in (mod Λ, Ab) is finitely generated if and only if there is an epimorphism $(M,\ )\to F\to 0$ with M in mod Λ, and is finitely presented if and only if there is an exact sequence $(N,\ )\to(M,\ )\to F\to 0$ with N, M in mod Λ.

Next, we recall that a functor F in (mod Λ, Ab) is said to be simple if $F \neq (0)$ and (0) and F are the only subfunctors of F. The pertinence of these notions to the question of whether or not Λ is of finite-representation type (i.e., mod Λ has only a finite number of nonisomorphic indecomposable objects) is given in the following proposition (see [2] for proof):

Proposition 1.1. Λ is of finite-representation type if and only if it satisfies

(a) every simple F in (mod Λ, Ab) is finitely presented, and
(b) every nonzero F in (mod Λ, Ab) has a simple subfunctor.

Thus it is of obvious interest to know when (mod Λ, Ab) satisfies (a) and (b). We begin with condition (a).

The following description of the simple functors F: mod $\Lambda \to$ Ab is given in [2]. The starting point is the observation that because an indecomposable object M of mod Λ has a local ring, $(M,\ )$ has a unique maximal subfunctor denoted by $\mathrm{r}(M,\ )$. Thus associated with each indecomposable object M in mod Λ is the simple functor $(M,\ )/\mathrm{r}(M,\ )$. Moreover, given any simple F in (mod Λ, Ab), there is a unique (up to isomorphism) indecomposable M in mod Λ such that $(M,\ )/\mathrm{r}(M,\ )\approx F$. Hence the correspondence $M\mapsto(M,\ )/\mathrm{r}(M,\ )$ gives a bijection between the isomorphism classes of simple objects in (mod Λ, Ab) and the isomorphism classes of indecomposable Λ-modules.

Another useful description of the simple objects in (mod Λ, Ab) is that F in (mod Λ, Ab) is simple if and only if there exists a unique (up to isomorphism) indecomposable M in mod Λ such that $F(M) \neq 0$, and if $F(M) \neq 0$, then $F(M)$ is a simple End(M)-module, where End(M) is the endomorphism ring of M.

From this discussion it is clear that every simple object in (mod Λ, Ab) is finitely generated, but it is still an open question for precisely which left artin rings Λ the simple objects in (mod Λ, Ab) are finitely presented. However it was shown in [4] that every simple object in (mod Λ, Ab) is finitely presented if Λ is an artin algebra. It is essentially for this reason that we are able to establish Theorem A only for artin algebras instead of left artin rings more generally. In particular we have, as an immediate consequence of Proposition 1.1., the following:

Proposition 1.2. An artin algebra Λ is of finite representation type if and only if every nonzero F in (mod Λ, Ab) has a simple subobject.

Returning to the case of an arbitrary left artin ring Λ, we recall (see [2]) that an F in (mod Λ, Ab) has a simple subobject if and only if there is an M in mod Λ such that there is an x in $F(M)$ that is universally minimal, i.e., $x \neq 0$ and a morphism $f\colon M \to N$ in mod Λ is a splittable monomorphism (there is $g\colon N \to M$ such that $gf = id_M$) if $F(f)(x)$ in $F(N)$ is not zero. Equivalently, we have

Proposition 1.3. F in (mod Λ, Ab) has no simple subobjects if and only if given any M in mod Λ and nonzero x in $F(M)$, there is a morphism $f\colon M \to N$ in mod Λ that is not a splittable monomorphism such that $F(f)(x)$ is not zero in $F(N)$.

We now show how to use F in (mod Λ, Ab) that have no simple subfunctors to create denumerably generated large indecomposable Λ-modules. This result will play a critical role in proving Theorems A and B as well as being of interest in its own right. Before doing this we need a few preliminaries.

Let F be in (mod Λ, Ab). Then, as we observed earlier, F has a projective presentation $\coprod_{i \in I} (N_i, \ ) \to \coprod_{i \in J} (M_j, \ ) \to F \to 0$ with the N_i and M_j in mod Λ. Associated with F is the functor $\tilde{F}$: Mod $\Lambda \to$ Ab obtained by viewing the representable functors $(N_i, \ )$ and $(M_j, \ )$ as representable functors on Mod Λ to Ab and defining $\tilde{F} = \operatorname{Coker}(\coprod_{i \in I} (N_i, \ ) \to \coprod_{j \in J} (M_j, \ ))$. In other words $\tilde{F}$ has the property that

$$\coprod_{i \in I} (N_i, X) \to \coprod_{j \in J} (M_j, X) \to \tilde{F}(X) \to 0$$

is exact for all X in Mod Λ. Since the N_i and M_j are finitely presented

Λ-modules, the functors $(N_j,\ \ )$, $(M_i,\ \ )$: Mod $\Lambda \to$ Ab commute with filtered direct limits. Consequently, the $\coprod_{j \in J}(N_j,\ \ )$, $\coprod_{i \in I}(M_i,\ \ )$: Mod $\Lambda \to$ Ab also commute with filtered direct limits, which implies that $\tilde{F} = \operatorname{Coker}(\coprod_{j \in J}(N_j,\ \ ) \to \coprod_{i \in I}(M_i,\ \ ))$ commutes with filtered direct limits. Therefore $\tilde{F}$: Mod $\Lambda \to$ Ab has the properties (a) $\tilde{F} \mid \text{mod } \Lambda = F$ and (b) $\tilde{F}$ commutes with filtered direct limits. Since every X in Mod Λ is the direct limit of its finitely generated submodules, these properties uniquely determine $\tilde{F}$, up to isomorphism. To simplify notation we will sometimes denote $\tilde{F}$ by F.

The other preliminary we need is the notion of a minimal element (see [2]). Let G: Mod $\Lambda \to$ Ab be an arbitrary functor. An element x in $G(X)$ said to be minimal if (a) $x \neq 0$ and (b) if $f: X \to X''$ is a proper epimorphism (an epimorphism that is not an isomorphism), then $G(f)(x)$ in $G(X'')$ is the zero element. It is now not difficult to establish the following properties of minimal elements (see [2]).

Proposition 1.4. Let G: Mod $\Lambda \to$ Ab be an arbitrary functor.

(a) A nonzero element x in $G(X)$ is minimal if and only if a morphism $f: X \to Y$ in Mod Λ is a monomorphism whenever it has the property $G(f)(x)$ in $G(Y)$ is not zero.

(b) If x in $G(X)$ is minimal, then X is an indecomposable Λ-module.

(c) If M is in mod Λ and x is a nonzero element of $G(M)$, then there is an epimorphism $f: M \to M''$ such that $G(f)(x)$ in $G(M'')$ is minimal.

We can now prove the main result of this section.

Theorem 1.5. Let F: mod $\Lambda \to$ Ab be a nonzero functor that has no simple subfunctors. Suppose M is in mod Λ and x is a nonzero element in $F(M)$. Then there exists a sequence of morphisms

$$M \xrightarrow{f_0} M_0 \xrightarrow{f_1} M_1 \xrightarrow{f_2} M_2 \longrightarrow \cdots \longrightarrow M_i \xrightarrow{f_{i+1}} M_{i+1} \longrightarrow \cdots$$

in mod Λ with the following properties:

(a) For each $i \geq 0$, the element $F(f_i \cdots f_1 f_0)(x) = x_i$ in $F(M_i)$ is a minimal element.

(b) Each f_i: $M_{i-1} \to M_i$ is a proper monomorphism, i.e., a monomorphism that is not an isomorphism, for $i \geq 1$ and f_0: $M \to M_0$ is an epimorphism.

(c) M_i is an indecomposable module in mod Λ for all $i \geq 0$.

(d) $\varinjlim M_i = N$ is a denumerably generated large indecomposable Λ-module.

Proof. Suppose $x \in F(M)$ is not zero. Then by Proposition 1.4 we know there is an epimorphism f_0: $M \to M_0$ such that $F(f_0)(x) = x_0$ is minimal in $F(M_0)$. Suppose we have defined, for $k \geq 0$, the sequence of $M \xrightarrow{f_0} M_0 \xrightarrow{f_1} \cdots \xrightarrow{f_k} M_k$ morphisms in mod Λ satisfying the conditions $F(f_i \cdots f_0)(x) = x_i$ is minimal in $F(M_i)$ and f_i: $M_i \to M_{i+1}$ is a proper monomorphism for all $i \leq 0, \ldots, k-1$. We now define f_{k+1}: $M_k \to M_{k+1}$ in mod Λ such that $F(f_{k+1} f \cdots f_0)(x) = x_{k+1}$ is a minimal element in $F(M_{k+1})$ and $f_{k+1}: M_k \to M_{k+1}$ is a proper monomorphism.

Since x_k is a minimal element in $F(M_k)$ we know that it is not zero. Because F has no universally minimal elements, since it has no simple subfunctors (see Proposition 1.3), it follows that there is a morphism g: $M_k \to L$ in mod Λ that is not a splittable monomorphism such that $F(g)(x_k)$ is a nonzero element of $F(L)$ (see Proposition 1.3). Thus by Proposition 1.4, there is an epimorphism h: $L \to M_{k+1}$ in mod Λ such that $x_{k+1} = F(h)(F(g)(x_k))$ in $F(M_{k+1})$ is a minimal element.

Let f_{k+1}: $M_k \to M_{k+1}$ be the composition hg. Since $x_{k+1} = F(f_{k+1})(x_k)$ is minimal and thus not zero, the fact that x_k is minimal in $F(M_k)$ implies that f_{k+1}: $M_k \to M_{k+1}$ is a monomorphism. Further, f_{k+1} must be a proper monomorphism since it is the composition $M_k \xrightarrow{g} L \xrightarrow{h} M_{k+1}$, with g not a splittable monomorphism. Thus the sequence of morphisms $M \xrightarrow{f_0} M_0 \xrightarrow{f_1} M_1 \longrightarrow \cdots \longrightarrow M_k \xrightarrow{f_{k+1}} M_{k+1}$ in mod Λ has the properties (a) for each i such that $0 \leq i \leq k+1$, the element $x_i = F(f_i f_{i-1} \cdots f_0)(x)$ is a minimal element in $F(M_i)$, and (b) f_0: $M \to M_0$ is an epimorphism while f_{i+1}: $M_i \to M_{i+1}$ is a proper monomorphism for $1 \leq i \leq k+1$. Therefore by induction we obtain a sequence of morphisms $M \xrightarrow{f_0} M_0 \xrightarrow{f_1} M_1 \longrightarrow \cdots \longrightarrow M_2 \xrightarrow{f_{i+1}} M_{i+1} \longrightarrow \cdots$, satisfying (a) and (b) of the theorem.

(c) follows from the fact that each x_i in $F(M_i)$ is minimal for $i \geq 0$ (see Proposition 1.4).

For proof of (d) assume now that $\tilde{F}$: Mod $\Lambda \to$ Ab is the unique extension of F: mod $\Lambda \to$ Ab, which commutes with filtered direct limits. Let $N = \varinjlim M_i$ and let g_i: $M_i \to N$ be the canonical morphisms that are obviously monomorphisms. Since $\tilde{F}(N) = \varinjlim \tilde{F}(M_i)$ and $F(f_{i+1})(x_i) = x_{i+1}$ for all $i \geq 0$, it follows that the $\tilde{F}(g_i)(x_i)$ are the same element y of $\tilde{F}(N)$ for all $i \geq 0$. We now show that N is indecomposable by showing that y in $\tilde{F}(N)$ is a minimal element.

Since none of the x_i in $\tilde{F}(M_i)$ is zero, the fact that $\tilde{F}$ commutes with filtered direct limits implies that y in $\tilde{F}(N)$ is not zero. Now suppose that N' is a nonzero submodule of N. Because $N = \varinjlim M_i$, it follows that there is an i such that $g_i^{-1}(N') \neq 0$. Thus we obtain the commutative diagram

$$\begin{array}{ccc} M_i & \xrightarrow{g_i} & N \\ \downarrow{\scriptstyle h_i} & & \downarrow{\scriptstyle h} \\ M_i/g_i^{-1}(N') & \xrightarrow{l} & N/N' \end{array}$$

where the h_i, h, l are the canonical morphisms. From this it follows that $\tilde{F}(h)(g) = \tilde{F}(lh_i)(x_i) = \tilde{F}(l)(\tilde{F}(h_i)(x_i))$. Since h_i is a proper epimorphism and x_i is a minimal element in $F(M_i)$, we know that $\tilde{F}(h_i)(x_i) = 0$. Hence $\tilde{F}(h)(y) = \tilde{F}(l)(\tilde{F}(h_i)(x_i)) = 0$. Since this is true for any nonzero submodule N' of N, it follows that y in $\tilde{F}(N)$ is minimal. Hence N is indecomposable.

Therefore to complete the proof of the theorem it only remains to show that N is a denumerably generated large Λ-module. But this follows easily from the fact that each $f_{k+1}: N_k \to N_{k+1}$ is a proper monomorphism in mod Λ.

As an immediate consequence of Theorem 1.5 we have

Corollary 1.6. If not every nonzero functor in (mod Λ, Ab) contains a simple subfunctor, then there are denumerably large indecomposable Λ-modules.

2. Proof of Theorem A

This section is devoted to proving Theorem A. Throughout this section Λ is a left artin ring and unless stated to the contrary, all Λ-modules are left Λ-modules.

We begin with

Proposition 2.1. The following statements are equivalent if the simple objects in (mod Λ, Ab) are finitely presented, such as for instance if Λ is an artin algebra.

(a) Λ is of finite-representation type.

(b) Every Λ-module is a sum of finitely generated Λ-modules.

(c) Every nonzero Λ-module has a finitely generated indecomposable summand.

(d) If $M_0 \xrightarrow{f_0} M_1 \xrightarrow{f_1} \cdots \longrightarrow M_i \xrightarrow{f_i} M_{i+1} \longrightarrow \cdots$ is a sequence of monomorphisms of indecomposable modules in mod Λ, then there is an integer n such that f_i is an isomorphism for all $i \geq n$.

(e) Every indecomposable Λ-module is finitely generated.

Proof. (a) implies (b). Shown in [2] and [6].

(b) implies (c). Obvious.

(c) implies (d). Suppose $M_0 \xrightarrow{f_0} M_1 \longrightarrow \cdots \longrightarrow M_i \xrightarrow{f_i} M_{i+1} \longrightarrow \cdots$ is a sequence of monomorphisms of indecomposable modules in mod Λ. Let $M = \varinjlim M_i$. Then the usual canonical morphisms g_i: $M_i \to M$ are monomorphisms that we will consider inclusions. Since $M \neq 0$, there is by (c) a finitely generated indecomposable summand M' of M. The fact that M' is finitely generated and $M = \bigcup_{i \in I} M_i$ implies that $M' \subset M_n$ for some n.

Hence $M' \subset M_i$ for all $i \geq n$. Since M' is a summand of M, it is a summand of any submodule of M containing it. In particular, M' is a summand of M_i for all $i \geq n$. This implies $M' = M_i$ for $i \geq n$ since the M_i are indecomposable and $M' \neq (0)$. Thus $M_i = M_n$ for all $i \geq n$, our desired result.

(d) implies (a). Since we are assuming that each simple object in (mod Λ, Ab) is finitely presented, in order to show that Λ is of finite representation type it suffices to show that (d) implies that each nonzero object in (mod Λ, Ab) has a simple subfunctor (see Proposition 1.1). Suppose there is a nonzero F in (mod Λ, Ab) that has no simple subfunctors. Since F is not zero, there is an M in mod Λ such that $F(M) \neq 0$. Then by Theorem 1.5 we know that there is a sequence of morphisms $M \xrightarrow{f_0} M_0 \xrightarrow{f_1} \cdots \longrightarrow M_i \xrightarrow{f_{i+1}} M_{i+1} \longrightarrow \cdots$ in mod Λ where the f_i are proper monomorphisms of indecomposable Λ-modules for all $i \geq 1$. This contradicts the hypothesis of (d). Hence every nonzero F in (mod Λ, Ab) has a simple subfunctor, which completes the proof that (d) implies (a) and the equivalence of (a) through (d).

(a) equivalent to (e). Since it is obvious that (a) implies (e), we only have to show that (e) implies (a). Again we only have to show that every nonzero object in (mod Λ, Ab) has a simple subobject. But we have already seen in Corollary 1.6 that if not every object in (mod Λ, Ab) has a simple subobject, then there are denumerably generated large indecomposable Λ-modules. This contradicts the hypothesis of (e) that every indecomposable Λ-module is finitely generated. Thus we have shown that (e) implies (a). This finishes the proof of the equivalence of (a) and (e) as well as the proof of Proposition 2.1.

Our next step in proving Theorem A is to show that an artin algebra Λ is of finite representation type if and only if given any sequence

$$\cdots \xrightarrow{f_{i+1}} M_i \xrightarrow{f_i} M_{i-1} \longrightarrow \cdots \longrightarrow M_1 \xrightarrow{f_0} M_0$$

of epimorphisms of indecomposable modules in mod Λ, there is an integer n such that if f_i is an isomorphism for all $i \geq n$.

Since Λ is an artin algebra, it is clear that Λ^{op}, the opposite ring of Λ, is also an artin algebra. Further we know that mod Λ and mod Λ^{op} are dual categories (see [4]). Thus mod Λ satisfies the above condition on epimorphisms between indecomposable objects if and only if for each sequence

$$N_0 \xrightarrow{f_0} N_1 \longrightarrow \cdots \longrightarrow N_i \xrightarrow{f_i} N_{i+1} \longrightarrow \cdots$$

of monomorphisms of indecomposables in mod Λ^{op}, there is an integer n such that f_i is an isomorphism for all $i \geq n$. But by 2.1 this condition is equivalent to Λ^{op} being of finite-representation type, since Λ^{op} is an artin algebra. Finally the duality between mod Λ and mod Λ^{op} shows that Λ^{op} is of finite-representation type if and only if Λ is of finite representation. We now obtain our desired result by putting these equivalences together.

The rest of Theorem A now follows from the duality between mod Λ and mod Λ^{op}.

We end this section with another application of Proposition 2.1 to the problem of determining when an arbitrary left artin ring Λ is of finite representation type. Before giving this application we establish some preliminary results.

Proposition 2.2. Let F be a finitely presented object of (mod Λ, Ab). Then its extension $\tilde{F}$: Mod $\Lambda \to$ Ab commutes with arbitrary products.

Proof. Since F is finitely presented, we know there is an exact sequence $(N, \ ) \to (M, \ ) \to \tilde{F} \to 0$ with N, M in mod Λ. Then we have the commutative exact diagram

$$\begin{array}{ccccccc}
0 & & 0 & & & & \\
\downarrow & & \downarrow & & & & \\
(N, \prod_{i\in I} X_i) & \to & (M, \prod_{i\in I} X_i) & \to & \tilde{F}(\prod_{i\in I} X_i) & \to & 0 \\
\downarrow & & \downarrow & & \downarrow & & \\
\prod_{i\in I} (N, X_i) & \to & \prod_{i\in I} (M, X_i) & \to & \prod_{i\in I} \tilde{F}(X_i) & \to & 0 \\
\downarrow & & \downarrow & & & & \\
0 & & 0 & & & &
\end{array}$$

since representable functors commute with products. Hence the right-hand vertical morphism is an isomorphism, which establishes the proposition.

Corollary 2.3. Let M be an indecomposable module in mod Λ such that the simple functor $F = (M, \ )/\mathrm{r}(M, \ )$ is finitely presented. Suppose $\{M_i\}_{i\in I}$ is a family of indecomposable modules in mod Λ such that M is a summand of $\prod_{i\in I} M_i$. Then $M \cong M_i$ for some i in I.

Proof. Since $F = (M, \ )/\mathrm{r}(M, \ )$ we know that if X is an indecomposable module in mod Λ, then $F(X) \neq 0$ if and only if $X \cong M$ (see Section 1). Suppose M is a summand of $\prod_{i\in I} X_i$ with the X_i indecomposable modules in mod Λ. Then because F is finitely presented we know by Proposition 2.2 that $F(\prod_{i\in I} X_i) = \prod_{i\in I} F(X_i)$. Since M is a summand of $\prod_{i\in I} X_i$ and $F(M) \neq 0$, it follows that $\prod_{i\in I} F(X_i) = F(\prod_{i\in I} X_i)$ is not zero. Hence $F(X_i) \neq (0)$ for some i in I, which implies that $X_i \approx M$, our desired result.

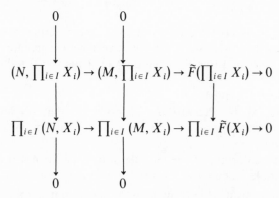

We now apply these results to establish the following

Proposition 2.4. For an arbitrary left artin ring, the following statements are equivalent.

(a) Λ is of finite representation type.
(b) Λ has the following properties:
  (i) Every left Λ-module is a sum of indecomposable modules in mod Λ;
  (ii) If M and $\{M\}_{i \in I}$ are indecomposable modules in mod Λ and M is a summand of $\prod M_i$, then $M \approx M_i$ for some i in I.
(c) Λ has the following properties:
  (i) Same as part (i) of (b);
  (ii) If $\{M_i\}_{i \in I}$ and $\{N_j\}_{j \in J}$ are families of indecomposable Λ-modules in mod Λ such that $\prod_{i \in I} M_i \cong \coprod_{j \in J} N_j$, then each N_j is isomorphic to an M_i for some i in I.

Proof. (a) implies (b). Since Λ is of finite representation type, we know that (b(i)) is satisfied. Also we know that every simple functor in (mod Λ, Ab) is finitely presented (see Proposition 1.1). Thus (b(ii)) follows trivially from Corollary 2.3.

(b) implies (c). Trivial.

(c) implies (a). By Proposition 2.1, to show that (c) implies (a) it suffices to show that (c) implies that every simple functor in (mod Λ, Ab) is finitely presented.

Let M be an indecomposable module in mod Λ. Then it was shown in [2] that the simple functor $(M,\ \ )/\mathrm{r}(M,\ \ )$ is finitely presented if and only if there is a morphism $M \xrightarrow{f} N$ in mod Λ that is not a splittable monomorphism such that, given any morphism $u\colon M \to X$ in mod Λ with X indecomposable that is not an isomorphism, there is a morphism $p\colon N \to X$ such that $u = pf$. We now show that (c) implies such a morphism $f\colon M \to N$ exists.

Let $\{M_i\}_{i \in I}$ be a complete set of nonisomorphic indecomposable objects in mod Λ, none of which is isomorphic to M. For each i in I, let $h_i\colon M \to M_i^{(M, M_i)}$ be the morphism defined by $h_i(m) = \{f(m)\}_{f \in (M, M_i)}$ and let $h\colon M \to \prod_{i \in I} M_i^{(M, M_i)}$ be the morphism induced by the h_i. Suppose we are given an indecomposable X in mod Λ and a morphism $u\colon M \to X$ that is not an isomorphism. Claim that there is a morphism $v\colon \prod M_i^{(M, M_i)} \to X$ such that $vh = u$.

If X is not isomorphic to M, then there is an isomorphism $s\colon M_i \to X$ for some i. Hence there is a morphism $t\colon M \to M_i$ such that $u = st$. Thus if we let $p\colon \prod_{i \in I} M_i^{(M, M_i)} \to X$ be the projection $\prod_{i \in I} M_i^{(M, M_i)} \to M_i^t$ composed with the isomorphism $s\colon M_i \to X$, we have $ph = u$.

Suppose $X \cong M$. Then $\operatorname{Im} u \neq X$ since u is not an isomorphism. Thus $\operatorname{Im} u$ is isomorphic to a finite product $\prod_{j \in J} X_j$ of indecomposable modules where each X_j is isomorphic to some M_i. For each $k \in J$ let u_k: $M \to X_k$ be the composition $M \to \prod_{j \in J} X_j \to X_k$. Then by our previous argument we know there is for each $k \in J$ a morphism p_k: $\prod_{i \in I} M_i^{(M, M_i)} \to X_k$ such that $u_k = p_k h$. Thus there is a p: $\prod_{i \in I} M_i^{(M, M_i)} \to \prod_{j \in J} X_j$ such that ph is the usual morphism $M \to \operatorname{Im} u$ given by u: $M \to X$. Therefore it follows that $ph = u$.

Now by (c) we know that $\prod_{i \in I} M_i^{(M, M_i)} \cong \coprod_{k \in K} L_k$ and each $L_k \cong M_i$ for some i in I. Since no M_i is isomorphic M, no L_k is isomorphic to M. Thus the morphism h: $M \to \coprod_{k \in K} L_k$ is not a splittable monomorphism, since the endomorphism rings of the M and L_k are local. Because M is finitely generated we know that there is a finite subset K' of K such that $\operatorname{Im} h \subset \coprod_{k \in K'} L_k$. Let f: $M \to \coprod_{k \in K'} L_k$ be the morphism defined by $f(m) = h(m)$ for all m in M. The fact that h is not a splittable monomorphism implies that $f: M \to \coprod_{k \in K'} L_k$ is not a splittable monomorphism. Also, since given any u: $M \to X$ between indecomposables in mod Λ that is not an isomorphism, there is a morphism p: $\coprod_{k \in K} L_k \to X$ such that $ph = n$, we have $h = (p \mid \coprod_{k \in K'} L_k) f$. Thus if we let $N = \coprod_{k \in K'} L_k$, then $f: M \to N$ is a morphism in mod Λ that is not a splittable monomorphism having the property that given any nonisomorphism u: $M \to X$ between indecomposable modules in mod Λ, there is a morphism p: $N \to X$ such that $u = pf$.

Thus (c) implies that for each indecomposable M in mod Λ, the simple functor $(M, \ )/\mathrm{r}(M, \ )$ is finitely presented. Since every simple functor in (mod Λ, Ab) is of the form $(M, \ )/\mathrm{r}(M, \ )$ for some indecomposable M in mod Λ, we have our desired result that (c) implies that the simple objects in (mod Λ, Ab) are finitely presented, finishing the proof that (c) implies (a).

These results suggest the following questions:

(1) For which left artin rings is it true that if M and $\{M_i\}_{i \in I}$ are indecomposables in mod Λ such that M is a summand of $\prod_{i \in I} M_i$, then $M \cong M_i$ for some i in I?

(2) For which left artin rings is it true that if $\{M_i\}_{i \in I}$ and $\{N_j\}_{j \in J}$ are families of indecomposables in mod Λ such that $\prod_{i \in I} M_i \cong \coprod_{j \in J} N_j$, then each N_j is isomorphic to some M_i?

(3) If Λ is a left artin ring such that every module is a sum of finitely generated indecomposable modules, does Λ satisfy the condition posed in Question (2)?

If the answers to (1) or (2) are all left artin rings, or if the answer to (3) is yes, then Proposition 2.4 shows that an arbitrary left artin ring Λ is of finite-representation type if and only if every left Λ-module is a sum of finitely generated indecomposable Λ-modules.

3. Proof of Theorem B

Throughout this section we assume that Λ is an artin algebra. Our aim is to prove Theorem B stated in the introduction. To this end, we must take a more careful look at the simple functors and the functors having no simple functors in (mod Λ, Ab).

In [3] it was shown that if M in mod Λ is indecomposable and not injective, then there is a monomorphism $f\colon M \to N$ in mod Λ that is not a splittable monomorphism having the property that given any morphism $g\colon M \to X$ in Mod Λ (i.e., X need not be finitely generated) that is not a splittable monomorphism, then there is a morphism $h\colon N \to X$ such that $hf = g$. From this it follows that (a) $(N,\ ) \to (M,\ ) \to (M,\ )/\mathrm{r}(M,\ ) \to 0$ is exact, and (b) the simple functor $F = (M,\ )/\mathrm{r}(M,\ )$ has the property that if X is in Mod Λ, then $F(X) \neq 0$ if and only if M is isomorphic to a summand of X.

Assertion (a) follows from [2].

To see (b), let X be an arbitrary Λ-module. By Yoneda's lemma, we know that $F(X) \neq (0)$ if and only if there is a nonzero morphism $(X,\ ) \to F$. Since $(X,\ )$ is a projective functor we know that

$$((X,\ ), (N,\ )) \to ((X,\ ), (M,\ )) \to ((X,\ ), F) \to 0$$

is exact. Hence there is a nonzero morphism $(X,\ ) \to F$ if and only if there is a morphism $(X,\ ) \to (M,\ )$ that cannot be "lifted" to $(N,\ )$. Or, stated differently, $F(X) \neq (0)$ if and only if there is a morphism $g\colon M \to X$ such that there is no $h\colon X \to N$ such that $hf = g$. Since only splittable monomorphisms $g\colon M \to X$ have this property, it follows that $F(X) \neq (0)$ if and only if M is a summand of X.

Suppose now that M is an injective indecomposable module in mod Λ. Then the socle S of M is simple. It is not difficult to check that the morphism $f\colon M \to M/S$, which is not a splittable monomorphism, has the property that if $g\colon M \to X$ in Mod Λ is not a splittable monomorphism, there is a morphism $h\colon M/S \to X$ such that $g = hf$. As above, it follows that $(M/S,\ ) \to (M,\ ) \to (M,\ )/\mathrm{r}(M,\ ) \to 0$ is exact and the simple functor $F = (M,\ )/\mathrm{r}(M,\ )$ has the property $F(X) \neq 0$ if and only if M is a summand of X.

Summarizing we have

Proposition 3.1. Let M be an indecomposable Λ-module in mod Λ and let F be the simple functor $(M,\ )/\mathrm{r}(M,\ )$ in (mod Λ, Ab). Then

(a) F is finitely presented.

(b) If X is in Mod Λ, then $F(X) \neq (0)$ if and only if M is a summand of X.

As an almost immediate consequence of this result we have the following corollaries.

Corollary 3.2. Let M be an indecomposable Λ-module in mod Λ and $\{X_i\}_{i \in I}$ a family of arbitrary Λ-modules. If M is isomorphic to a summand of the product $\prod_{i \in I} X_i$, then M is isomorphic to a summand of X_i for some i.

Proof. Let F be the simple functor $(M, \ )/\mathrm{r}(M, \ )$. Since M is a summand of $\prod_{i \in I} X_i$, we have that $F(\prod_{i \in I} X_i) \neq 0$. But $F(\prod_{i \in I} X_i) = \prod_{i \in I} F(X_i)$ since F is finitely presented (see Proposition 2.2). Hence $F(X_i) \neq (0)$ for some i in I and so M is a summand of X_i by Proposition 3.1.

Of more immediate concern to us is

Corollary 3.3. Suppose X in Mod Λ has no finitely generated indecomposable summands. Then $((X, \ ), F) = 0$ for all F in (mod Λ, Ab) of finite length.

Proof. Since F is of finite length, there is a finite filtration $(0) = F_0 \subset F_1 \subset \cdots \subset F_n = F$ of F with the property F_{i+1}/F_i is a simple object of (mod Λ, Ab) for $i = 0, \ldots, n-1$. Hence we know by Proposition 3.1 that $((X, \ ), F_{i+1}/F_i) = 0$ for $i = 0, \ldots, n-1$. From this it follows trivially that $((X, \ ), F) = 0$.

The last preliminary result we need before proving Theorem B is the following:

Proposition 3.4. Let F be a finitely generated functor in (mod Λ, Ab). Then F has finite length if and only if there are only a finite number of nonisomorphic indecomposable M in mod Λ such that $F(M) \neq 0$.

Proof. In [2], it was shown that an F in (mod Λ, Ab) is of finite length if and only if (a) there is only a finite number of nonisomorphic indecomposable M in mod Λ such that $F(M) \neq 0$, and (b) $F(M)$ has finite length viewed as a module over End(M) for each indecomposable M in mod Λ. Hence if F has finite length there are only a finite number of nonisomorphic indecomposable M in mod Λ such that $F(M) \neq 0$.

Suppose F is a finitely generated functor in (mod Λ, Ab), i.e., there is an epimorphism $(N, \ ) \to F \to 0$ with N in mod Λ. Let C be the center of Λ. Since Λ is an artin algebra we know that Λ is a finitely generated C-module and C is an artin commutative ring. Hence for each M in mod Λ, we know that (N, M) is a finitely generated C-module. But $\text{End}(M)$ is also an artin algebra since it is a C-algebra that is a finitely generated C-module. Hence (N, M) is a finitely generated $\text{End}(M)$-module and hence of finite length. It follows that $F(M)$ is an $\text{End}(M)$ of finite length for each M in mod Λ since $(N, M) \to F(M) \to 0$ is an exact of $\text{End}(M)$-modules. Therefore if there are only a finite number of nonisomorphic indecomposable M in mod Λ such that $F(M) \neq 0$, it follows that F is of finite length since each $F(M)$ is a module of finite length over $\text{End}(M)$.

We now restate and prove Theorem B.

Theorem 3.5. The following statements are equivalent for an indecomposable Λ-module M in mod Λ.

(a) $(M, N) \neq 0$ for an infinite number of nonisomorphic indecomposable N in mod Λ.
(b) $(M, X) \neq (0)$ for some denumerably generated large indecomposable module X.
(c) $(M, X) \neq (0)$ for some X with no finitely generated indecomposable summands.

Proof. (a) implies (b). In view of Proposition 3.4, the hypothesis that $(M, N) \neq 0$ for an infinite number of nonisomorphic indecomposable N in mod Λ is the same as the hypothesis that $(M, \ )$ does not have finite length. Consequently the subfunctor $\text{l.f.}(M, \ ) \neq (M, \ )$, where $\text{l.f.}(M, \ )$ is the subfunctor of $(M, \ )$ generated by the subfunctors of $(M, \ )$ of finite length. Further, since the simple functors in (mod Λ, Ab) are finitely presented, $F = (M, \ )/\text{l.f.}(M, \ )$ has no simple subfunctors (see [2] for details). It now follows from Theorem 1.5 that because $F(M) \neq 0$, there is a nontrivial morphism $M \to X$ with X a denumerably generated large indecomposable Λ-module.

(b) implies (c). Trivial.

(c) implies (a). Suppose $f: M \to X$ is a nontrivial morphism with X a Λ-module having no finitely generated indecomposable summands. Then the morphism $(f, \ ): (X, \ ) \to (M, \ )$ is also not trivial. Hence by Corollary 3.3, $(M, \ )$ does not have a finite length. Consequently by Proposition 3.4 there must be an infinite number of nonisomorphic indecomposable N in mod Λ such that $(M, N) \neq 0$. This shows that (c) implies (a) and finishes the proof of the theorem.

We conclude this paper with some remarks concerning the following observation.

Let M be an indecomposable module in mod Λ. We have seen that there is a morphism $f: M \to N$ in mod Λ that is not a splittable monomorphism such that given any morphism $g: M \to X$ in Mod Λ that is not a splittable monomorphism, there is a morphism $h: N \to X$ such that $g = hf$.

As an easily verified consequence of this we have that a morphism $f: M \to X$ in Mod Λ is a splittable monomorphism if and only if for each finitely generated submodule X' of X containing $\operatorname{Im} f$, the induced morphism $M \to X'$ is a splittable monomorphism. Now this result concerning indecomposable M in mod Λ can be used to show directly that the statement remains true if one assumes only that M is an arbitrary module in mod Λ instead of an indecomposable one. Instead of doing this, we shall give an independent proof of a somewhat more general result.

Let C be the center of the artin algebra Λ and I the injective envelope over C of the C-module $C/\mathfrak{r}$ where $\mathfrak{r}$ is the radical of the artin ring C. Let D: Mod $\Lambda \to$ Mod Λ^{op} be the functor given by $X \to \operatorname{Hom}_C(X, I)$ for all X in Mod Λ. We define D: Mod $\Lambda^{\mathrm{op}} \to$ Mod Λ similarly. It is well known that X in Mod Λ is finitely generated if and only if $D(X)$ in Mod Λ^{op} is finitely generated and that the induced functor D: mod $\Lambda \to$ mod Λ^{op} is an equivalence of categories. The analogous results hold for D: Mod $\Lambda^{\mathrm{op}} \to$ Mod Λ.

Since each X in Mod Λ^{op} is a filtered direct limit of its finitely generated submodules, $D(X)$ is a filtered inverse limit of finitely generated Λ-modules. On the other hand, if Y in Mod Λ is a filtered inverse limit of finitely generated Λ-modules, then $Y \cong D(X)$ for some X in Mod Λ^{op}. For suppose $Y = \varprojlim Y_i$ with the Y_i in mod Λ. Then the filtered inverse limit system $\{Y_i\}_{i \in I}$ gives rise to the filtered direct limit system $\{D(Y_i)\}$ of finitely generated Λ^{op}-modules. If we let $X = \varinjlim D(Y_i)$, then $D(X) = \varprojlim Y_i = Y$, proving our assertion. Thus we have established

Lemma 3.6. Let X lie in Mod Λ. Then X is a filtered inverse limit of finitely generated Λ-modules if and only if there is a Y in Mod Λ^{op} such that $D(Y) \cong X$.

As a consequence of this we have

Proposition 3.7. Let X in Mod Λ be a filtered inverse limit of finitely generated Λ-modules. Then

(a) If $Z = \varinjlim Z_i$ in Mod Λ, then $\operatorname{Ext}_\Lambda{}^i(Z, X) \cong \varprojlim \operatorname{Ext}^i(Z_i, X)$ for all $i \geq 0$.

(b) If $0 \to X \xrightarrow{f} Y \xrightarrow{g} Z$ is an exact sequence in Mod Λ, then it is splittable if and only if for each finitely generated submodule Z_i of Z, the exact sequence $0 \to X \to g^{-1}(Z_i) \to Z_i \to 0$ is splittable.

Proof. (a) Since X is a filtered inverse limit of finitely generated Λ-modules, we know that $X = D(Y)$ for some Y in Mod Λ^{op}. Now it is well known (see [5]) that there are functorial isomorphisms

$$\operatorname{Ext}_\Lambda{}^i(Z, D(Y)) \cong \operatorname{Hom}_C(\operatorname{Tor}_i{}^\Lambda(Z, Y), I)$$

for all Z in Mod Λ and Y in Mod Λ^{op} and all $i \geq 0$. Since the $\mathrm{Tor}_i^\Lambda(\ ,Y)$ commutes with filtered direct limits, $\mathrm{Hom}_C(\mathrm{Tor}_i^\Lambda(\ ,Y), I) = \mathrm{Ext}^i(\ ,D(Y))$ takes filtered direct limits to filtered inverse limits, giving our desired result.

Part (b) is a ready consequence of (a).

We can now prove our final result.

Proposition 3.8. Suppose $f: X \to Y$ is a morphism in Mod Λ and X is a filtered inverse limit of finitely generated Λ-modules. Then the following statements are equivalent:

(a) $f: X \to Y$ is a splittable monomorphism.

(b) The morphism of functors $\otimes f$: $\otimes X \to \otimes Y$ is a monomorphism.

(c) For each submodule Y' or Y containing $\mathrm{Im}\, f$ such that $Y'/\mathrm{Im}\, f$ is finitely generated, the induced morphism $X \to Y'$ is a splittable monomorphism.

Proof. (a) implies (b). This is trivial.

(b) implies (c). Since $\otimes f: \otimes X \to \otimes Y$ is a monomorphism, it follows that f: $X \to Y$ is a monomorphism. Suppose $0 \to X \xrightarrow{f} Y \xrightarrow{g} Z \to 0$ is exact. Then for each finitely generated submodule Z', we obtain the exact sequence $0 \to X \to g^{-1}(Z') \to Z' \to 0$. Since $0 \to \otimes X \to \otimes Y \to \otimes Z \to 0$ is exact, it follows that $0 \to \otimes X \to \otimes g^{-1}(Z') \to \otimes Z' \to 0$ is exact for each finitely generated submodule Z' of Z. Since each Z' is finitely presented, it is well known that $0 \to \otimes X \to \otimes g^{-1}(Z') \to \otimes Z' \to 0$ being exact implies that $0 \to X \to g^{-1}(Z') \to Z' \to 0$ is a splittable exact sequence. Thus $0 \to X \to g^{-1}(Z') \to Z' \to 0$ is a splittable exact sequence for all finitely generated submodules Z' of Z, which shows that (b) implies (c).

(c) implies (a). Obviously the hypothesis of (d) implies that $f: X \to Y$ is a monomorphism with the property that if $0 \to X \xrightarrow{f} Y \xrightarrow{g} Z \to 0$ is exact, then $0 \to X \xrightarrow{f} g^{-1}(Z') \to Z' \to 0$ is a splittable monomorphism for all finitely generated submodules Z' of Z. This implies that since X is an inverse limit of finitely generated Λ-modules, that the exact sequence $0 \to X \to Y \to Z \to 0$ is splittable as we saw in Proposition 3.7(b). This shows that (c) implies (a) and completes the proof of the proposition.

REFERENCES

[1] Auslander, M., Representation theory of artin algebras I, *Comm. in Algebra* (1974), 177–268.

[2] Auslander, M., Representation theory of artin algebras II, *Comm. in Algebra* (1974), 269–310.

[3] Auslander, M., and Reiten, I., Representation theory of artin algebras III, *Comm. in Algebra* (1975), 239–294.

[4] Auslander, M., and Reiten, I., Stable equivalence of dualizing R-varieties, *Advances in Math.* (1974), 306–366.
[5] Cartan, H., and Eilenberg, S., "Homological Algebra." Princeton Univ. Press, Princeton, New Jersey, 1956.
[6] Ringel, C., and Tachikawa, H., *QF*-3 rings, *J. Reine Angew. Math.* (1975), 49–72.

Written with the partial support of NSF GP-33406X3.

AMS 16A46

DEPARTMENT OF MATHEMATICS
BRANDEIS UNIVERSITY
WALTHAM, MASSACHUSETTS

Reduced Bar Constructions on deRham Complexes

KUO-TSAI CHEN

To Professor Samuel Eilenberg with great admiration.

The purpose of this paper is to set up the reduced bar construction functorially on a deRham complex in a suitable way so that it may be paired with the cobar construction of a smooth chain complex. The associated Eilenberg–Moore spectral sequence of the reduced bar construction converges to the loop space cohomology in the simply connected case. Our other aim is to obtain integral homological information about loop spaces through the use of integral deRham cohomology classes.

In Section 1, the definition of the bar construction is slightly modified so that it may be defined on a deRham complex, which is not a connected DG algebra. In Section 2, the E_1 term of the associated Eilenberg–Moore spectral sequence is discussed. The bar construction is treated in such a way that a pairing with the cobar construction through iterated integration can be described in Section 3. By making use of results in [3], we show that, under reasonable conditions, the induced pairing for the Eilenberg–Moore spectral sequences are nondegenerate.

If A is the deRham complex in consideration and if $\overline{B}(A)$ is the reduced bar construction on A, then the E_1 term of the Eilenberg–Moore spectral sequence is isomorphic to $\overline{B}(H(A))$ as DG algebras. In Section 4, we determine the torsion-free quotient chain complex of the E^1 term of the spectral sequence of the cobar construction (on a smooth chain complex) through the cup products of the integral deRham cohomology classes. Applications are given in Section 5 for cases where either $E_1 = E_\infty$ or $E_2 = E_\infty$. Generally speaking, if $E_1 = E_\infty$ [i.e., all Masey products are zero in $H(A)$], then the loop space homology algebra is free. If $E_2 = E_\infty$, then the loop space homology is isomorphic to the homology of a DG-free (associative) algebra, whose differential can be given through the cup products in $H(A)$. As examples, we made the following computations.

(a) The integral homology $H_*(\Omega S^n)$, $n > 1$, is a polynomial ring of a single generator of degree $n-1$. (See Serre [13] and Bott and Samelson [2].)

(b) The real homology of the twice-iterated loop space $\Omega^2 S^n$, n odd, is reduced to the calculation of the homology of the DG-free algebra generated by $X_1, X_2, \ldots$ of respective degrees $n-2, 2n-3, \ldots$ such that $\partial X_1 = 0$ and, for $m > 1$, $\partial X_m = \sum \binom{m}{i} X_i X_{m-i}$.

(c) For an $(n-1)$-connected orientable compact C^∞ manifold of dimension $2n$, $n > 1$, the real loop space homology is isomorphic to the homology of a DG-free algebra generated by $X_1, \ldots, X_m, X$ with $\partial X_1 = \cdots = \partial X_m = 0$ and $\partial X = \sum c_{ij} X_i X_j$, where $\deg X_i = n-1$, $\deg X = 2n-1$, m is the nth Betti number, and the $m \times m$ integral matrix (c_{ij}) is given by cup products of cohomology classes.

The bar construction of Eilenberg and Mac Lane [7, 8] was originally used for chain complexes. A cochain complex version can be found in [14]. Our version as presented in Section 1 has been announced earlier.[1] The essence of the material in Section 2 can be found in the works of Eilenberg and Moore [9] and Smith [14, 15]. We also point out a relation between Sullivan's minimal models [10, 16] and the reduced bar construction, which seems to be relevant also to Quillen's rational homotopy theory [12]. The geometrical part of this work depends on the cobar construction of Adams [1].

1. Bar Construction on DG Algebras

Let k be a commutative ring with 1. For every element v of a graded k-module, define $Jv = (-1)^{\deg v} v$.

Let $A = \{A^p\}$ be a commutative DG k-algebra with a differential d of degree $+1$ and with $A^p = 0$ for $p < 0$. Denote by A^+ the DG A-module

[1] Notice of *Amer. Math. Soc.* **20** (1973), A-357.

obtained from A by replacing A^0 with 0. A special case of A in our mind is that of A being the deRham complex of a differentiable manifold. Therefore $\wedge$ is used to denote the multiplication in A. When $a \in A^0$ and $w \in A^p$, we may write $aw = a \wedge w$.

Let M and N be DG A-modules. Let $w, w_1, w_2, \ldots$ denote (homogeneous) elements of A^+, and let x and y denote elements of M and N, respectively.

Denote by $\overline{T}(A)$ the graded tensor algebra on A^+, and set

$$T(N, A, M) = N \otimes \overline{T}(A) \otimes M,$$

where $\otimes = \otimes_k$. Then $T(N, A, M) = \sum T^r(N, A, M)$ such that $T^0(N, A, M) = N \otimes M$ and, for $r < 0$, $T^r(N, A, M) = 0$ and

$$T^r(N, A, M) = N \otimes (\otimes^r A^+) \otimes M, \qquad r > 0.$$

Thus $T^r(N, A, M)$ is spanned by elements of the type $y[w_1|\cdots|w_r]x$. Define

$$\deg y[w_1|\cdots|w_r]x = \deg y + \sum_{1 \le i \le r} (-1 + \deg w_i) + \deg x.$$

The differential d is decomposed as a sum

$$d = \bar{d} + d' + d''$$

so that, if $u = [w_1|\cdots|w_r]$, then

$$\begin{aligned}\bar{d}(yux) = &\sum_{1 \le i \le r} (-1)^i Jy[Jw_1|\cdots|Jw_{i-1}|dw_i|w_{i+1}|\cdots|w_r]x \\ &- \sum_{1 \le i < r} (-1)^i Jy[Jw_1|\cdots|Jw_{i-1}|Jw_i \wedge w_{i+1}|w_{i+2}|\cdots|w_r]x;\end{aligned}$$

$$d'(yux) = dy[w_1|\cdots|w_r]x - Jy \wedge w_1[w_2|\cdots|w_r];$$

$$d''(yux) = (-1)^r Jy[Jw_1|\cdots|Jw_r]\, dx - (-1)^r Jy[Jw_1|\cdots|Jw_{r-1}]w_r \wedge x.$$

When $r = 0$, set $\bar{d}(y[\ ]x) = 0$, $d'(y[\ ]x) = dy[\ ]x$, and $d''(y[\ ]x) = Jy[\ ]\, dx$.

There is a filtration F of $T(N, A, M)$ with

$$F^s T(N, A, M) = \sum_{r \le s} T^r(N, A, M).$$

Thus $N \otimes M = F^0 T(N, A, M) \subset F^1 T(N, A, M) \subset \cdots$.

Observe that d', d'', and $\bar{d}$ are natural. Write $Ju = (-1)^r[Jw_1|\cdots|Jw_r]$. If $v = [w_{r+1}|\cdots|w_{r+s}]$, write

$$\begin{aligned}uv &= [w_1|\cdots|w_{r+s}], \\ u \wedge v &= [w_1|\cdots|w_{r-1}|w_r \wedge w_{r+1}|w_{r+2}|\cdots|w_{r+s}].\end{aligned}$$

Then

$$\bar{d}(yuvx) = y((\bar{d}u)v + Ju\bar{d}v - Ju \wedge v)x.$$

Let $f \in A^0$. For the sake of abbreviation, write $[\cdots|fw_i|\cdots] = [w_1|\cdots|w_{i-1}|fw_i|w_{i+1}|\cdots|w_r]$.

Let $\mathfrak{J}$ be the graded k-submodule of $T(N, A, M)$ spanned by all elements $yR_i(u, f)x$ where $x \in M$, $y \in N$, $f \in A^0$, $u = [w_1|\cdots|w_r]$, w's $\in A^+$, $r \geq 0$, $0 \leq i \leq r$, and

$$yR_1(u, f)x = -y[fw_1|\cdots]x + yf[w_1|\cdots|w_r]x + y[df|w_1|\cdots|w_r]x;$$
$$yR_i(u, f)x = -y[\cdots|fw_i|\cdots]x + y[\cdots|fw_{i-1}|\cdots]x + y[\cdots|w_{i-1}|df|w_i|\cdots]x, \quad 1 < i \leq r;$$
$$yR_0(u, f)x = y[w_1|\cdots|w_r]fx + y[\cdots|fw_r]x + y[w_1|\cdots|w_r|df]x.$$

Remark. In the case of $r = 0$, $R_0([\ ], f) = -[\ ]f + f[\ ] + [df]$.

Lemma. The submodule $\mathfrak{J}$ is closed under the differential d of $T(N, A, M)$.

Proof. We are going to give a proof for the case of A having an augmentation ε: $A \to k$ and of $M = N = k$ being the A-module induced by the augmentation. This special case will be adequate for subsequent applications in this paper.

For $T(k, A, k)$, we have $d' = d'' = 0$ and $d = \bar{d}$. We may set $x = y = 1$ and write $R_i(u, f) = yR_i(u, f)x$. Verify that

$$\bar{d}R_0([\ ], f) = 0;$$
$$\bar{d}R_0(w, f) = [dw]\varepsilon f - [df \wedge w + f\, dw] - [dw|df] + [Jw \wedge df] = -R_0([dw], f);$$
$$\bar{d}R_0(u, f) = \bar{d}([w_1|\cdots|w_{r-1}]R_0([w_r], f)) \in \mathfrak{J};$$
$$\bar{d}R_1([w], f) = [df \wedge w + f\, dw] - f[dw] - [df|dw] - [df \wedge w] = -R_1([dw], f);$$
$$\bar{d}R_1(u, f) = \bar{d}(R_1([w_1], f)[w_2|\cdots|w_r]) \in \mathfrak{J};$$
$$\begin{aligned}\bar{d}R_2([w_1|w_2], f) = {} & [dw_1|fw_2] - [Jw_1|df \wedge w_2 + f\, dw_2] \\ & - [Jw_1 \wedge fw_2] - [df \wedge w_1 + f\, dw_1|w_2] \\ & + [fJw_1|dw_2] + [fJw_1 \wedge w_2] - [dw_1|df|w_2] \\ & + [Jw_1|df|dw_2] + [Jw_1 \wedge df|w_2] + [Jw_1|df \wedge w_2] \\ = {} & R_2(-[dw_1|w_2] + [Jw_1|dw_2], f);\end{aligned}$$

and, for $2 \leq i \leq r$,

$$\bar{d}R_i(u, f) = \bar{d}([w_1|\cdots|w_{i-2}]R_2([w_{i-1}|w_i], f)[w_{i+1}|\cdots|w_r]) \in \mathfrak{J}.$$

Hence $d\mathfrak{J} \subset \mathfrak{J}$.

Define $B(N, A, M)$ to be the quotient complex $T(N, A, M)/\mathfrak{J}$. The differential of $B(N, A, M)$ is also denoted by d, and the induced filtration of $B(N, A, M)$ is also denoted by F.

Observe that, when A is connected, then $\mathfrak{J} = 0$ and $B(N, A, M) = T(N, A, M)$. In particular, $B(A, A, M)$ is the usual bar construction of M over A (see [14]) with some differences in the sign convention.

2. The Eilenberg–Moore Spectral Sequence

Hereafter k will be assumed to be a field. For the commutative DG k-algebra A, $H(A)$ will be assumed to be connected.

The filtered cochain complex $B(N, A, M)$ gives rise to a spectral sequence $\{E_r, d_r\}$, $r \geq 0$, namely, the Eilenberg–Moore spectral sequence.

Theorem. The E_1 term of the Eilenberg–Moore spectral sequence is isomorphic, as a cochain complex, to $B(H(N), H(A), H(M))$, where $H(A)$, $H(M)$, and $H(N)$ are equipped with trivial differentials. If $H^1(A) = 0$, then the spectral sequence converges to $H(B(N, A, M))$.

Proof. The convergence of the spectral sequence follows from the fact that $F^r B(H(N), H(A), H(M))$ consists of elements of degree $\geq r$.

In order to compute E_1, let $\overline{A}$ be a connected DG subalgebra of A such that $\overline{A}^p = A^p$ for $p > 1$ and

$$A^1 = dA^0 \oplus \overline{A}^1.$$

We shall shortly verify that the inclusion $\overline{A} \subset A$ induces an isomorphism

$$T(N, \overline{A}, M) = B(N, \overline{A}, M) \approx B(N, A, M). \tag{2.1}$$

From this isomorphism, we observe that

$$d_0 \colon E_0{}^s \approx T^s(N, \overline{A}, M) \to T^s(N, \overline{A}, M) \approx E_0{}^s$$

is given by

$$\begin{aligned} y[w_1|\cdots|w_s]x \mapsto{} & dy[w_1|\cdots|w_s]x \\ & + \sum_{1 \leq i \leq s} (-1)^i Jy[Jw_1|\cdots|Jw_{i-1}|dw_i|w_{i+1}|\cdots|w_s]x \\ & + (-1)^s Jy[Jw_1|\cdots|Jw_s]\, dx. \end{aligned}$$

In other words, d_0 is the differential of the tensor product $N \otimes (\otimes^s \overline{A}^+) \otimes M$ with the degree of $\overline{A}^+$ being lowered by 1. By the Künneth formula,

$$E_1{}^s \approx H(N) \otimes (\otimes^s H(\overline{A}^+)) \otimes H(M),$$

with the degree of $H(\bar{A}^+)$ being lowered by 1, i.e.,

$$E_1 \approx B(H(N), H(A), H(M)).$$

Moreover $d_1: E_1{}^s \to E_1^{s-1}$ is given by

$$\begin{aligned} \bar{y}[\bar{w}_1|\cdots|\bar{w}_s]\bar{x} \mapsto & - J\bar{y} \wedge \bar{w}_1[\bar{w}_2|\cdots|\bar{w}_s]\bar{x} \\ & - \sum_{1\le i<s} (-1)^i J\bar{y}[J\bar{w}_1|\cdots|J\bar{w}_{i-1}|J\bar{w}_i \wedge \bar{w}_{i+1}|\bar{w}_{i+2}|\cdots|\bar{w}_s]\bar{x} \\ & - (-1)^s J\bar{y}[J\bar{w}_1|\cdots|J\bar{w}_{s-1}]\bar{w}_s \wedge \bar{x}, \end{aligned}$$

where $\bar{x}$, $\bar{y}$, $\bar{w}_i$ are cohomology classes of $x \in M$, $y \in N$, and $w_i \in \bar{A}^+$, respectively.

It remains to verify the isomorphism (2.1). Let

$$\begin{aligned} U^r = N \otimes \{ & dA^0 \otimes (\otimes^r A^+) + A^+ \otimes dA^0 \otimes (\otimes^{r-1} A^+) + \cdots \\ & + (\otimes^r A^+) \otimes dA^0\} \otimes M, \end{aligned}$$

and $U = \sum \oplus U^r$. Then

$$T(N, A, M) = T(N, \bar{A}, M) \oplus U$$

and U has a filtration induced by that of $T(N, A, M)$. Let the exact sequence

$$U \xrightarrow{\alpha} T(N, A, M) \to B(N, A, M) \to 0 \tag{2.2}$$

be such that α is given by

$$y[w_1|\cdots|w_{i-1}|df|w_i|\cdots|w_r]x \mapsto yR_i(u, f)x$$

where $u = [w_1|\cdots|w_r]$ and $f \in A^0$ are as given in Section 1. The filtration of $T(N, A, M)$ gives rise to an exact sequence

$$0 \longrightarrow \operatorname{gr} U \xrightarrow{\operatorname{gr} \alpha} \operatorname{gr} T(N, A, M) \longrightarrow \operatorname{gr} B(N, A, M) \longrightarrow 0$$

where

$$\operatorname{gr} \alpha: U = \operatorname{gr} U \to \operatorname{gr} T(N, A, M) = T(N, A, M)$$

is simply the inclusion. Hence

$$\operatorname{gr} B(N, A, M) \approx T(N, \bar{A}, M) = \operatorname{gr} T(N, \bar{A}, M),$$

and the theorem is proved.

Corollary 1. Under the same hypothesis as the theorem,

$$E_2 \approx \operatorname{Tor}_{H(A)}(H(N), H(M)).$$

Definition. For a commutative DGA k-algebra A, define the reduced bar construction of A to be

$$\overline{B}(A) = B(k, A, k),$$

where k is taken as a two-sided DG A-module via the augmentation of A.

Corollary 2. If a morphism of commutative DGA k-algebras $\hat{A} \to A$ induces an isomorphism $H(\hat{A}) \approx H(A)$ with $H^0(A) \approx k$, then there are induced isomorphisms of the Eilenberg–Moore spectral sequences

$$E_r(\overline{B}(\hat{A})) \approx E_r(\overline{B}(A)), \qquad r \geq 1.$$

Material in this section is mostly modifications of known results. In particular, this corollary can be traced back to Eilenberg and Mac Lane [7].

The same corollary implies that, if $\rho\colon M \to A$ is a Sullivan's minimal model for A, then

$$H(\overline{B}(M)) \approx H(\overline{B}(A)).$$

A main point of the minimal model is that one can read the indecomposibles of $\mathrm{Tor}_A(k, k) = H(\overline{B}(A))$ during the construction of M when A is a deRham complex of a CW complex equipped with a suitable differentiable structure (see [10]).

3. A Pairing of $\overline{B}(A)$ and a Cobar Construction

The field k will be henceforth that of the real or complex numbers. Let X be a path-connected differentiable space as defined in [3]. (The word "differentiable" should be replaced more appropriately by "predifferentiable.") Choose a base point $x_0 \in X$. Then the deRham complex $\Lambda(X)$ has an augmentation arising from the base point. Let A be a DGA subalgebra of $\Lambda(X)$.

Denote by ΩX the piecewise smooth loop space of X at x_0. In [3], we have defined a DG subalgebra A' of $\Lambda(\Omega X)$ that is spanned by iterated integrals of the type $\int w_1 \cdots w_r$, $w_i \in A^+$, $r \geq 0$. There is a filtration

$$k = A'(0) \subset A'(1) \subset A'(2) \subset \cdots$$

of A' such that $A'(s)$ is spanned by iterated integrals of the type $\int w_1 \cdots w_r$, $r \leq s$, $w_i \in A^+$. Because of the formulas (4.1.2)–(4.1.4) in [3], there is a cochain map

$$\overline{B}(A) \to A' \tag{3.1}$$

given by $[w_1|\cdots|w_r] \mapsto \int w_1 \cdots w_r$. This map respects the filtrations and is, as a matter of fact, a morphism of DG algebras.

Let $\hat{C}_*(X)$ be the chain complex of smooth simplices of X with all vertices at x_0. In [3], Section 4.4, the differentiable space X is assumed to be simply connected so that we may further demand $\hat{C}_1(X) = 0$. For the nonsimply connected case, we need to modify the treatment in [3], Section 4.6, so that if σ is a 1-simplex in $\hat{C}_*(X)$, then $\bar{\sigma}$ is the 0-chain $c_\sigma - c_{x_0}$, where c_σ and c_{x_0} are respectively the 0-cubes at the loop $\sigma \in \Omega X$ and the constant loop at x_0.

Denote by $F(\hat{C}_*)$ the cobar construction on $\hat{C}_*(X)$, which has a filtration

$$F(\hat{C}_*) = F_0(\hat{C}_*) \supset F_1(\hat{C}_*) \supset \cdots$$

such that $F_s(\hat{C}_*)$ is spanned by all $[c_1|\cdots|c_r]$, $r \le s$, where the c are simplices of $\hat{C}_*(X)$ of positive degrees. Let $\{E^r, \partial^r\}$ be the associated Eilenberg–Moore spectral sequence.

We shall assume that the deRham theorem holds for A. By this we mean that

(a) The canonical map from $\hat{C}_*(X)$ into the normalized singular chain complex of X is a chain equivalence.

(b) $H(A) \approx H^*(X; k)$ via integration on $\hat{C}_*(X)$.

We shall also assume that $H_*(X)$ is of finite type. Under the above assumptions, we establish a pairing

$$A' \times F(\hat{C}_*) \to k$$

through integration. This pairing respects the filtrations and induces a nondegenerate pairing

$$E_1 \times E^1 \otimes k \to k,$$

where $\{E_r, d_r\} = \{E_r(\overline{B}(A)), d_r\}$ is the Eilenberg–Moore spectral sequence of the reduced bar construction $\overline{B}(A)$.

We are now led to the next assertion by the chain map (3.1).

Theorem. Let X be a path-connected differentiable space, whose singular homology is of finite type. Let the deRham theorem hold for a DG subalgebra A of $\Lambda(X)$, which is augmented through a choice of a base point of X. Then there is a pairing of a filtered cochain complex and a filtered chain complex

$$\overline{B}(A) \times F(\hat{C}_*) \to k, \tag{3.2}$$

which induces nondegenerate pairings

$$E_r(\overline{B}(A)) \times E^r \otimes k \to k, \qquad r \ge 1. \tag{3.3}$$

A theorem of Adams [1] yields the next assertion.

Corollary 1. Under the same hypothesis as in the theorem, if X is a simply connected topological space, then $H^*(\bar{B}(A)) \approx H^*(\Omega X; k)$.

For the nonsimply connected case, we mention the next result, which depends on the work [5]. (See also [4].)

Corollary 2. Under the same hypothesis as in the theorem, if the fundamental group $G = \pi_1(X, x_0)$ is finitely generated and if $\varepsilon: kG \to k$ is the augmentation of the group algebra kG, then $H^0(\bar{B}(A))$ is naturally isomorphic with the Hopf algebra of all k-valued linear functionals on kG that annul some power of the augmentation ideal ker ε.

It should be pointed out here that there is an independent result of Sullivan in terms of his minimal models, which can be shown to be equivalent to Corollary 2 for the case of X being a CW complex.

4. The E_1 Term of the Cobar Construction

The reduced bar construction $\bar{B}(A)$ has a Hopf algebra structure. The multiplication is given by

$$[w_1|\cdots|w_r][w_{r+1}|\cdots|w_{r+s}] = \sum \varepsilon(\sigma, p_1, \ldots, p_{r+s})[w_{\sigma(1)}|\cdots|w_{\sigma(r+s)}]$$

summing over all (r, s)-shuffles σ, where $p_i = -1 + \deg w_i$ and $\varepsilon(\sigma, p_1, \ldots, p_{r+s}) = \pm 1$ is as described in [3], Section 4.1. The comultiplication is given by

$$[w_1|\cdots|w_r] \mapsto \sum_{0 \le i \le r} [w_1|\cdots|w_i] \otimes [w_{i+1}|\cdots|w_r].$$

Owing to [3] (1.6.2), this comultiplication corresponds to the multiplication of $F(\hat{C}_*)$ via the pairing (3.2). As a matter of fact, the map (3.1) is a morphism of DG k-algebras.

Similarly $E_1(\bar{B}(A)) \approx \bar{B}(H(A))$ has a DG k-algebra structure, which can be seen more easily through its dual $\bar{B}(H(A))^*$ as follows:

Let $\{\bar{w}_1, \bar{w}_2, \ldots\}$ be a basis of $H^+(A)$ with

$$\deg \bar{w}_1 \le \deg \bar{w}_2 \le \cdots.$$

Let $X_1, X_2, \ldots$ be a dual basis for $(H^+(A))^*$. Write

$$J\bar{w}_j \wedge \bar{w}_k = \sum c^i_{jk} \bar{w}_i, \qquad c^i_{jk} \in k.$$

Then, as a Hopf algebra, $\bar{B}(H(A))^*$ can be taken as the free k-algebra generated by $X_1, X_2, \ldots$ with the degrees of the X being lowered by one, i.e.,

$$\deg X_i = -1 + \deg \bar{w}_i.$$

The comultiplication of $\overline{B}(H(A))^*$ is given by

$$X_i \mapsto X_i \otimes 1 + 1 \otimes X_i.$$

The differential of $\overline{B}(H(A))^*$ is given by

$$X_i \mapsto \sum c^i_{jk} X_j X_k.$$

A straightforward verification shows that $\overline{B}(H(A))^*$, as described above, is indeed the dual DG Hopf k-algebra of $\overline{B}(H(A))$. Also observe that

$$J\overline{w}_k \wedge \overline{w}_j = -(-1)^{(-1+\deg \overline{w}_j)(-1+\deg \overline{w}_k)} J\overline{w}_j \wedge \overline{w}_k$$

so that $c^i_{kj} = -(-1)^{(\deg X_j)(\deg X_k)} c^i_{jk}$. If [,] denotes the Lie bracket for graded algebras, then

$$\sum c^i_{jk} X_j X_k = \tfrac{1}{2} \sum c^i_{jk} [X_j, X_k].$$

The next assertion is a consequence of the preceding theorem.

Theorem. Let X be a path-connected differentiable space, whose singular homology is of finite type. Choose a base point $x_0 \in X$. Assume that the deRham theorem holds for a DG subalgebra A of $\Lambda(X)$. Let $\overline{w}_1, \overline{w}_2, \ldots$ with ascending degrees be a basis for $H^+(A)$, and let

$$J\overline{w}_j \wedge \overline{w}_k = \sum c^i_{jk} \overline{w}_i.$$

Let $\{E^r, \partial^r\}$ be the Eilenberg–Moore spectral sequence of the cobar construction $F(\hat{C}_*)$. Then $E^1 \otimes k$ is isomorphic to the free k-algebra $k[X_1, X_2, \ldots]$ generated by $X_1, X_2, \ldots$ with $\deg X_i = -1 + \deg \overline{w}_i$, and the differential ∂^1 is given by

$$X_i \mapsto \tfrac{1}{2} \sum c^i_{jk} [X_j, X_k].$$

Let $H(A; Z)$ denote the ring of integral cohomology classes in $H(A)$. If $\{\overline{w}_1, \overline{w}_2, \ldots\}$ is a basis for $H^+(A; Z)$, then all c^i_{jk} are integers. The free ring $Z[X_1, X_2, \ldots]$, as a subring of $k[X_1, X_2, \ldots]$, is closed under the differential.

Corollary. Under the same hypothesis as the theorem, if $\{\overline{w}_1, \overline{w}_2, \ldots\}$ is an integral basis of $H^+(A)$, then the torsion free quotient of E^1 is isomorphic to the free ring $Z[X_1, X_2, \ldots]$, the differential ∂^1 is given by

$$X_i \mapsto \sum c^i_{jk} X_j X_k.$$

Remark. If $H_*(X)$ is torsion free, then it follows from the Künneth formula that E^1 is also torsion free and is therefore isomorphic to $Z[X_1, X_2, \ldots]$ as chain complexes. We have also

$$\mathrm{Hom}_Z(E^1, Z) \approx B_Z(H(A; Z)), \tag{4.1}$$

which is the integral reduced bar construction on the graded ring $H(A; Z)$. If, furthermore, $H(A; Z)$ is a graded polynomial ring (with all odd-degree elements vanishing) on $v_1, v_2, \ldots$, then $\mathrm{Tor}_{H(A;Z)}(Z, Z)$ is isomorphic to the exterior algebra on generators $u_1, u_2, \ldots$ over Z with $\deg u_i = -1 + \deg v_i$. This is a known result, which follows from a Koszul resolution.

5. Applications

Consider first a differentiable space X with $E_1 = E_\infty$ (i.e., all Massey products in $H(A)$ being zero). We have

$$H(\bar{B}(A)) \approx E_1 \approx \bar{B}(H(A)).$$

If X is simply connected, then

$$H^*(\Omega X; k) \approx \bar{B}(H(A)),$$

and there is an isomorphism of algebras

$$H_*(\Omega X; k) \approx k[X_1, X_2, \ldots].$$

If, furthermore, $H_*(X)$ is torsion free, then

$$H_*(\Omega X) \approx Z[X_1, X_2, \ldots],$$

and, on the cohomological level,

$$H^*(\Omega X) \approx \bar{B}_Z(H(A; Z)),$$

where $\bar{B}_Z(H(A; Z))$ denotes the reduced bar construction over Z on the integral cohomology ring $H(A; Z) \subset H(A)$.

Example 1. Let w be the normalized volume element of S^n, $n > 1$, and let A be the DG subalgebra of $\Lambda(S^n)$ having $\{1, w\}$ as a basis. Then $\bar{B}_Z(H(A; Z))$ has as a basis

$$[\ ],\quad [w],\quad [w|w], \ldots.$$

Let A' be DG subalgebra of $\Lambda(\Omega S^n)$ having a basis consisting of iterated integrals

$$1,\quad \int w,\quad \int ww, \ldots. \tag{5.1}$$

These iterated integrals represent a basis for the integral cohomology classes of ΩS^n. We also obtain the known fact that $H_*(\Omega S^n)$ is isomorphic to the polynomial ring of a single generator of degree $n - 1$.

When X is simply connected with $E_2 = E_\infty$, then

$$H^*(\Omega X; k) \approx H(\bar{B}(H(A))) = \mathrm{Tor}_{H(A)}(k, k)$$

and

$$H_*(\Omega X; k) \approx H(k[X_1, X_2, \ldots]).$$

If, furthermore, both $H_*(X)$ and E^2 (or $\mathrm{Tor}_{H(A;\,Z)}(Z, Z)$) are torsion free, then

$$H_*(\Omega X) \approx H(Z[X_1, X_2, \ldots]).$$

Many differentiable spaces have the property $E_2 = E_\infty$. Among them are two important classes of differentiable manifolds:

(a) If the DG subalgebra A of $\Lambda(X)$ is such that $dA = 0$, then $H(A) = A$ and $E_1 = \overline{B}(A)$. For $r > 1$, each differential d_r is induced by d_1 and is therefore trivial. This class of manifolds includes compact Lie groups and, more generally, compact symmetric Riemannian manifolds.

(b) If X is a compact Kähler manifold with $A = \Lambda(X)$, then the exterior differential can be written as a sum $d = d' + d''$ owing to the complex structure of X. In [11], Griffiths *et al.* have shown that there are chain equivalences of complexes:

$$\{A, d\} \leftarrow \{\mathrm{Ker}\ d', d\} \rightarrow \{H(A), \text{trivial differential}\}.$$

According to Section 4,

$$E_1 = E_1(\overline{B}(A)) \approx E_1(\overline{B}(H(A))) = \overline{B}(H(A)).$$

Observe that $E_2(\overline{B}(H(A))) = E_\infty(\overline{B}(H(A)))$. Hence $E_2 = E_\infty$.

Before giving concrete examples of the case of $E_2 = E_\infty$, we take note that, if $H_*(X)$ is torsion free and if $H(A; Z)$ is a polynomial ring, then, according to (4.1), the cohomology ring $H^*(\Omega X)$ is isomorphic to an exterior algebra over Z on generators $u_1, u_2, \ldots$.

Example 2. Let $X = \Omega S^n$, $n \geq 3$ being odd. Let A' be the DG subalgebra of $\Lambda(\Omega S^n)$ having as a basis the iterated integrals

$$1, \quad u_1 = \int w, \quad u_2 = \int ww, \ldots$$

as given in (5.1). Then the deRham theorem holds for A' on ΩS^n, and $dA' = 0$. According to [3], (4.1.1),

$$u_r \wedge u_s = \binom{r+s}{r} u_{r+s}$$

where $\binom{r+s}{r}$ is the binomial coefficient. Thus $H_*(\Omega^2 S^n; k)$ is isomorphic to the homology of the DG free algebra $k[X_1, X_2, \ldots]$ with $\deg X_m = m(n-1) - 1$ and the differential ∂ such that $\partial X_1 = 0$ and, for $m > 1$,

$$\partial X_m = \sum_{1 \leq i < m} \binom{m}{i} X_i X_{m-i}.$$

Example 3. Let X be an $(n-1)$-connected $2n$-dimensional orientable compact C^∞ manifold, $n > 1$. Let $w_1, \ldots, w_m$ be closed n-forms such that $\overline{w}_1, \ldots, \overline{w}_m$ is an integral basis for $H^n(X)$. Write $J\overline{w}_j \wedge \overline{w}_k = c_{jk}\overline{w}$, where w is a normalized volume element of X. Let N be a subspace of $\Lambda^{2n}(X)$ such that $kw \oplus N = \Lambda^{2n}(X)$. Choose $(2n-1)$-forms w_{jk} such that

$$Jw_j \wedge w_k + dw_{jk} = c_{jk} w$$

and $dw_{jk} \in N$. Let A be the DG subalgebra of $\Lambda(X)$ spanned by 1, w, w_i, w_{ij}, where $i, j = 1, \ldots, m$. Then the map

$$A \to H(A)$$

given by $w_i \mapsto \overline{w}_i$, $w_{ij} \mapsto 0$, $w \mapsto \overline{w}$, and $dw_{ij} \mapsto 0$, is a well-defined chain map, which is a homology isomorphism. Thus $E_2 = E_\infty$. Let $1, X_1, \ldots, X_m, \overline{X}$ be the dual basis for $H(A)^*$. Then $H_*(\Omega X)$ is isomorphic to the DG-free algebra $k[X_1, X_2, \ldots]$, whose differential ∂ is given by

$$\partial X_i = 0, \qquad \partial \overline{X} = \sum c_{ij} X_i X_j.$$

REFERENCES

[1] J. F. Adams, On the cobar construction, *Colloque de topologie algebrique, Louvain* (1956), 81–87.

[2] R. Bott and H. Samelson, On the Pontrjagin product in spaces of paths, *Comment. Math. Helv.* **27** (1953), 320–337.

[3] K. T. Chen, Iterated integrals of differential forms and loop space homology, *Ann. of Math.* **97** (1973), 217–246.

[4] K. T. Chen, Fundamental groups, nilmanifolds and iterated integrals, *Bull. Amer. Math. Soc.* **79** (1973), 1033–1035.

[5] K. T. Chen, Iterated integrals, fundamental groups and covering spaces, *Trans. Amer. Math. Soc.* **206** (1975), 83–98.

[6] K. T. Chen, Connections, holonomy and path space homology, *Proc. Symposia Pure Math., Amer. Math. Soc.* **27** (1975), 39–52.

[7] S. Eilenberg and S. Mac Lane, On the groups of $H(\Pi, n)$, I, *Ann. of Math.* **58** (1953), 55–106.

[8] S. Eilenberg and S. Mac Lane, On the groups of $H(\Pi, n)$, II, *Ann. of Math.* **60** (1954), 49–139.

[9] S. Eilenberg and J. C. Moore, Homology and fibrations I, *Comment. Math. Helv.* **40** (1966), 199–236.

[10] E. Fridelander, P. A. Griffiths, and J. Morgan, "Homotopy Theory and Differential Forms," Seminario di Geometria, Firenze (1972), (mimeographed).

[11] P. Deligne, P. A. Griffiths, J. Morgan, and D. Sullivan, Real homotopy theory of Kähler manifolds, *Inventiones Math.* **29** (1975), 245–274.

[12] D. G. Quillen, Rational homotopy theory, *Ann. of Math.* **90** (1969), 205–295.

[13] J. -P. Serre, Homologie singuliere des espaces fibres, *Applications Ann. of Math.* **54** (1951), 425–505.

[14] L. Smith, Homological algebra and the Eilenberg–Moore spectral sequence, *Trans. Amer. Math. Soc.* **129** (1967), 58–93.
[15] L. Smith, "Lectures on the Eilenberg–Moore Spectral Sequence." Springer-Verlag, Berlin, 1970.
[16] D. Sullivan, Differential forms and the topology of manifolds, *Proc. Conf. on Manifolds, Tokyo, 1973.*

Work supported in part by the National Science Foundation under NSF-GP-34257.

AMS 53C65, 55H20, 57A65.

DEPARTMENT OF MATHEMATICS
UNIVERSITY OF ILLINOIS AT URBANA-CHAMPAIGN
URBANA, ILLINOIS

Flatness and Projectivity of Modules That Come from $\mathbb{C}$-Sets

CHARLES CHING-AN CHENG and BARRY MITCHELL

Laudal [5] has characterized those small categories $\mathbb{C}$ relative to which the (inverse) limit functor

$$\lim: \mathrm{Ab}^{\mathbb{C}} \to \mathrm{Ab}$$

is exact. The analogous problem for the colimit functor was treated by Isbell and Mitchell [4]. In the former case the problem amounted to determining when $\Delta\mathbb{Z}$, the constant functor at $\mathbb{Z}$, is projective when considered as a module over the ringoid $\mathbb{Z}\mathbb{C}$, and in the latter case one had to determine when this module is flat. Now $\Delta\mathbb{Z}$ is a special case of a module that "comes from" a $\mathbb{C}$-set. That is to say, starting with a $\mathbb{C}$-set M (covariant functor $M: \mathbb{C} \to \mathrm{Sets}$), one can form the module (additive functor) $\mathbb{Z}M$: $\mathbb{Z}\mathbb{C} \to \mathrm{Ab}$ whose value at C is the free abelian group on MC. One can then consider the more general problem of determining when this module is flat or projective. In each case the results obtained are in terms of the comma category (Y, M) where $Y: \mathbb{C}^{\mathrm{op}} \to \mathrm{Sets}^{\mathbb{C}}$ is the Yoneda imbedding. If one considers nonadditive flatness and projectivity of M, the results are simple and well-known and are included here only for completeness. However in the additive case the matter is not so simple, and one needs hypotheses on M or $\mathbb{C}$. Sometimes the results are valid for arbitrary coefficient rings in place of $\mathbb{Z}$, and this leads to questions of when projective modules over ringoids of the form $K\mathbb{C}$ (K a field) are free, or in other words, coproducts of representables. Some instances where this is the case are mentioned at the end, but the reader is advised that the conjecture which this gives rise to is to be accepted with a grain of salt.

1. Nonadditive Flatness

Let $\mathbb{C}$ be a small category, $M \in \mathrm{Sets}^{\mathbb{C}^{\mathrm{op}}}$, and $N \in \mathrm{Sets}^{\mathbb{C}}$. The *tensor product* of the $\mathbb{C}^{\mathrm{op}}$-set M with the $\mathbb{C}$-set N is defined as

$$M \times_{\mathbb{C}} N = \bigcup_{A \in |\mathbb{C}|} MA \times NA/\sim,$$

where the union is disjoint, and where $\sim$ is the equivalence relation generated by $(x\alpha, y) \sim (x, \alpha y)$, where say, $x \in MB$, $\alpha \in \mathbb{C}(A, B)$, $y \in NA$. If X is a set, then we have an obvious bijection

$$\mathrm{Sets}(M \times_{\mathbb{C}} N, X) \simeq \mathrm{Sets}^{\mathbb{C}}(N, \mathrm{Sets}(M, X)),$$

which is natural in all three variables. Thus $M \times_{\mathbb{C}}$ is a left adjoint and so preserves all colimits. We also have the natural isomorphism

$$M \times_{\mathbb{C}} \mathbb{C}(A, \ ) \simeq MA, \tag{1}$$

which sends (x, α) to $x\alpha$.

If $T: \mathscr{A} \to \mathscr{B}$ is a functor and $B \in |\mathscr{B}|$, then we let (T, B) denote the category whose objects are morphisms $TA \to B$ and whose morphisms are commutative triangles

$$\begin{array}{ccc} TA & \longrightarrow & TA' \\ & \searrow \quad \swarrow & \\ & B & \end{array}$$

In particular, if $M \in \mathrm{Sets}^{\mathbb{C}^{\mathrm{op}}}$, and $Y: \mathbb{C} \to \mathrm{Sets}^{\mathbb{C}^{\mathrm{op}}}$ denotes the Yoneda imbedding, then by the Yoneda lemma the objects of (Y, M) can be identified with elements x of M, and a morphism from x to y can be identified with a pair (y, α) such that $y\alpha = x$. Then one checks that

$$M = \mathrm{colim}(Y, M) \xrightarrow{\pi} \mathbb{C} \xrightarrow{Y} \mathrm{Sets}^{\mathbb{C}^{\mathrm{op}}},$$

where π is the natural projection.

Recall that a functor is *left exact* if it preserves finite limits. Recall also that a category $\mathbb{C}$ is *filtered* if

I. every pair of objects map to a common object, and

II. for every pair of morphisms α, α' with common domain and codomain, there is a morphism β which *filters* them; that is, $\beta\alpha = \beta\alpha'$.

It is easy to see that the components of a category $\mathbb{C}$ are filtered if $\mathbb{C}$ satisfies II and if every pair of arrows with a common domain can be completed to a not necessarily commutative square. Of course II then guarantees that such a square can be made commutative.

Proposition 1.1. $M \times_{\mathbb{C}}$ is left exact if and only if (Y, M) is filtered.

Proof. If (Y, M) is filtered, then M is a filtered colimit of representables. Since tensoring with representables preserves finite limits by (1), and since filtered colimits in Sets preserve finite limits, it follows that $M \times_{\mathbb{C}}$ is left exact.

Now suppose that $M \times_{\mathbb{C}}$ is left exact. If $x \in MA$ and $x' \in MA'$, then since the natural map

$$M \times_{\mathbb{C}}[\mathbb{C}(A, \ ) \times \mathbb{C}(A', \ )] \to MA \times MA'$$

is a bijection, (x, x') is the image of, say, $(y, (\alpha, \alpha'))$. Hence $y\alpha = x$ and $y\alpha' = x'$, and so y is an object of (Y, M) to which the two objects x, x' map.

Now suppose that we have $y\alpha = x = y\alpha'$. Let E be the equalizer in $\mathrm{Sets}^{\mathbb{C}}$ of $\mathbb{C}(\alpha, \ )$ and $\mathbb{C}(\alpha', \ )$. Then

$$\begin{array}{ccccc} M \times_{\mathbb{C}} E \to & M \times_{\mathbb{C}} \mathbb{C}(B, \ ) & \rightrightarrows & M \times_{\mathbb{C}} \mathbb{C}(A, \ ) \\ & \wr\| & & \wr\| \\ & MB & & MA \end{array}$$

is an equalizer. Therefore some $(z, \beta) \in M \times_{\mathbb{C}} E$ must go to $y \in MB$. Hence $z\beta = y$ and $\beta\alpha = \beta\alpha'$. This shows that any pair of morphisms with common domain and codomain in (Y, M) can be filtered, and so (Y, M) is filtered.

Remark. For a generalization of the proposition in topos theory, see Diaconescu [3].

2. Flatness of $\mathbb{Z}M$

A *ringoid* is a small preadditive category $\mathscr{C}$. A *left $\mathscr{C}$-module* is a covariant additive functor $L: \mathscr{C} \to \mathrm{Ab}$. The group of natural transformations from L to L' is denoted by $\mathrm{Hom}_{\mathscr{C}}(L, L')$, and the category of left $\mathscr{C}$-modules is denoted by $\mathrm{Mod}\,\mathscr{C}$. For information on how to generalize module theory over rings to module theory over ringoids, we refer the reader to [7]. In particular, if $E \in \mathrm{Mod}\,\mathscr{C}^{\mathrm{op}}$ and $L \in \mathrm{Mod}\,\mathscr{C}$, then the tensor product

$E \otimes_{\mathscr{C}} L$ is defined similarly to its nonadditive counterpart of Section 1, and flatness of a module and purity of a monomorphism are defined as usual. If $0 \to E' \xrightarrow{u} E \to E'' \to 0$ is exact and E'' is flat, then u is pure. The following criterion for purity, due to P. M. Cohn, is proved by imitating the usual proof for modules over a ring [1, Section 2].

Proposition 2.1. Let E' be a submodule of the right $\mathscr{C}$-module E. In order that the inclusion be pure, it is necessary and sufficient that for each family of equations of the form

$$a_i' = \sum_{j \in J} b_j \alpha_{ij}, \qquad i \in I,$$

with I and J finite, $a_i' \in E'A_i$, $b_j \in EB_j$, and $\alpha_{ij} \in \mathscr{C}(A_i, B_j)$, there is a family $b_j' \in E'B_j$ such that

$$a_i' = \sum_{j \in J} b_j' \alpha_{ij}, \qquad i \in I.$$

We recall from [4] that the *affinization* of a small category $\mathbb{C}$ is the subcategory aff $\mathbb{C}$ of the additivization $\mathbb{Z}\mathbb{C}$ consisting of those morphisms whose integer coefficients sum to one. If $N \in \mathrm{Ab}^{\mathbb{C}}$, then N can be regarded as a left $\mathbb{Z}\mathbb{C}$-module, and hence by restriction, as an object of $\mathrm{Ab}^{\mathrm{aff}\,\mathbb{C}}$. Then it is easy to see that

$$\mathrm{colim}_{\mathbb{C}}\, N = \mathrm{colim}_{\mathrm{aff}\,\mathbb{C}}\, N.$$

It follows that if E is a right $\mathscr{C}$-module which is a colimit of a diagram over $\mathbb{C}$ of flat modules, and if aff $\mathbb{C}$ has filtered components, then E is flat.

In particular, consider a $\mathbb{C}^{\mathrm{op}}$-set M. Then composition of M with the free abelian group functor Sets $\to$ Ab yields a $\mathbb{Z}\mathbb{C}^{\mathrm{op}}$-module $\mathbb{Z}M$. If $x \in MC$, we denote C by $|x|$. Then we have

$$M = \operatorname*{colim}_{x \in (Y, M)} \mathbb{C}(\ , |x|),$$

and so since the free abelian group functor preserves colimits (being a left adjoint), we have

$$\mathbb{Z}M = \operatorname*{colim}_{x \in (Y, M)} \mathbb{Z}\mathbb{C}(\ , |x|).$$

It follows that if aff(Y, M) has filtered components, then $\mathbb{Z}M$ is flat. Under an assumption on M we shall prove the converse.

Theorem 2.2. If M is a $\mathbb{C}^{\mathrm{op}}$-set such that $M\alpha$ is an injection for all $\alpha \in \mathbb{C}$, then $\mathbb{Z}M$ is flat if and only if aff(Y, M) has filtered components.

Proof. We have seen the "if" direction without any assumption on M.

Now consider the exact sequence of $\mathbb{Z}\mathbb{C}^{\mathrm{op}}$-modules

$$0 \to K \xrightarrow{u} \bigoplus_{x \in M} \mathbb{Z}\mathbb{C}(\ ,|x|) \xrightarrow{\varepsilon} \mathbb{Z}M \to 0$$

where $\varepsilon(1_{|x|}) = x$, and so KB consists of all linear combinations $\sum r_i(x_i, \beta_i)$ such that $\sum r_i x_i \beta_i = 0$ in $\mathbb{Z}M(B)$. Here the r_i are integers, and the symbol (x, β) represents the morphism β considered as an element of $\mathbb{C}(\ ,|x|)$. Consider a diagram in (Y, M)

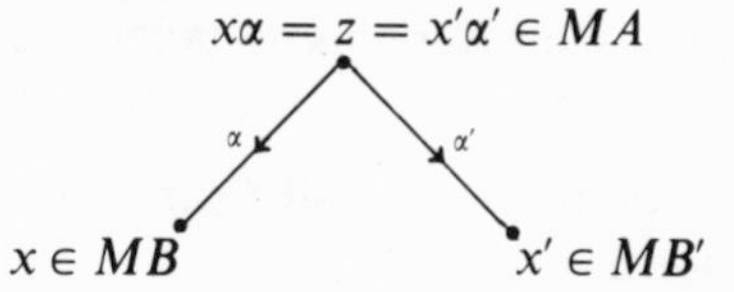

(1)

Then $(x,\alpha) - (x',\alpha') \in KA$. If $\mathbb{Z}M$ is flat, then u is pure, and so by Proposition 2.1 we can write

$$(x,\alpha) - (x',\alpha') = \sum_{i=1}^{n} r_i(y_i, \beta_i)\alpha - \sum_{i=1}^{n'} r_i'(y_i', \beta_i')\alpha' \tag{2}$$

with

$$\sum r_i y_i \beta_i = 0 = \sum r_i' y_i' \beta_i'. \tag{3}$$

Now if we take only the terms on the right of (2) with $y_i\beta_i\alpha = z$ and $y_i'\beta_i'\alpha' = z$, then the equation is still valid. But then by the assumption on M, these must be precisely the terms such that $y_i\beta_i = x$ and $y_i'\beta_i' = x'$, and so (3) will still be true. Then rewriting (2), we find

$$(x,\alpha) - \sum_{i=1}^{n} r_i(y_i, \beta_i\alpha) = (x',\alpha') - \sum_{i=1}^{n'} r_i'(y_i', \beta_i'\alpha').$$

Since the sum of coefficients on either side is one, we must have a basis element on one side equal to a basis element on the other, which shows that (1) can be completed to a commutative square in (Y, M).

Now consider a diagram in aff(Y, M)

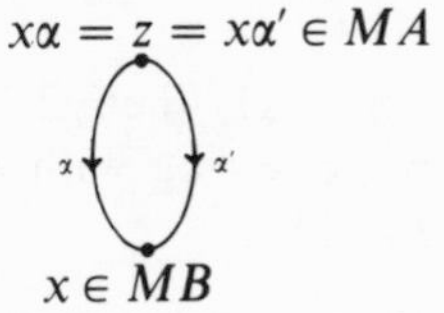

Then $(x, 1)(\alpha - \alpha') \in KA$, and so again by Proposition 2.1 we can write

$$(x, 1)(\alpha - \alpha') = \sum_{i=1}^{n} r_i(y_i, \beta_i)(\alpha - \alpha'), \tag{4}$$

where $\sum r_i y_i \beta_i = 0$. Again we may assume that $y_i \beta_i = x$ for all i using the assumption on M. Rewriting (4), we obtain

$$\left[(x, 1) - \sum_{i=1}^{n} r_i(y_i, \beta_i)\right](\alpha - \alpha') = 0.$$

Thus a linear combination of morphisms in (Y, M) with coefficients summing to one filters α and α'. But these morphisms do not necessarily have a common codomain, and so their linear combination does not comprise a morphism of aff(Y, M). However this is taken care of by the following lemma.

Lemma 2.3. The components of aff $\mathbb{C}$ are filtered if and only if every diagram in $\mathbb{C}$ of the form

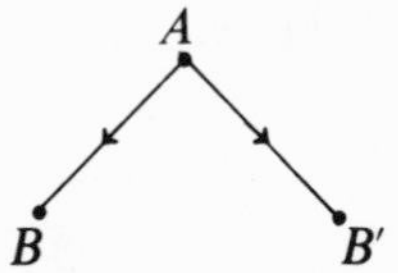

can be completed to a (not necessarily commutative) square in $\mathbb{C}$, and for each diagram of aff $\mathbb{C}$ of the form

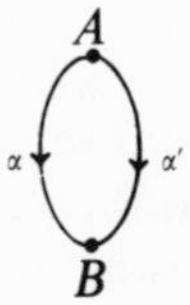

there is an element β of $\bigoplus_{C \in |\mathbb{C}|} \mathbb{Z}\mathbb{C}(B, C)$ with coefficients summing to one such that $\beta\alpha = \beta'\alpha'$.

Proof. If $\beta = \sum r_i \beta_i$ and $\beta\alpha = \beta\alpha'$, then using the first assumption we can choose, for each object in $\mathbb{C}$ that appears as the codomain of a β_i, a morphism γ with common codomain. Composing each β_i with the appropriate γ does not alter the validity of $\beta\alpha = \beta\alpha'$, but now β will be a morphism of aff $\mathbb{C}$.

If we take M to be $\mathbb{C}^{op}$-set $\Delta 1$, then $\mathbb{Z}M$ is $\Delta\mathbb{Z}$. Since the colimit functor $\mathrm{Ab}^{\mathbb{C}} \to \mathrm{Ab}$ is given by tensoring with $\Delta\mathbb{Z}$, and since $(Y, \Delta 1) = \mathbb{C}$, we obtain the following result of [4].

Corollary 2.4. If $\mathbb{C}$ is any small category, then colim: $\mathrm{Ab}^{\mathbb{C}} \to \mathrm{Ab}$ is exact if and only if the components of aff $\mathbb{C}$ are filtered.

Let α be a morphism of a ringoid $\mathscr{C}$ which is an epimorphism (left cancellation). Then $\mathscr{C}(\alpha, C)$ is an injection for all C. Hence $F(\alpha)$ is an injection for all free modules F (that is, coproducts of representables). Since every flat module is a direct limit of free modules (see [6] or [8]), it follows that $E(\alpha)$ is an injection for all flat modules E. Now if α is an epimorphism of the small, nonadditive category $\mathbb{C}$, α is also an epimorphism of $\mathbb{Z}\mathbb{C}$. Hence if $\mathbb{Z}M$ is flat, then $\mathbb{Z}M(\alpha)$ is an injection, and so $M(\alpha)$ is an injection. Thus from the theorem we obtain

Corollary 2.5. If all morphisms of $\mathbb{C}$ are epimorphisms, then $\mathbb{Z}M$ is flat if and only if $\text{aff}(Y, M)$ has filtered components.

Corollary 2.6. If $U:\mathbb{C} \to \mathbb{D}$ takes all morphisms to epimorphisms, then the left adjoint of $\text{Ab}^{\mathbb{D}} \to \text{Ab}^{\mathbb{C}}$ is exact if and only if $\text{aff}(U, D)$ has filtered components for all $D \in |\mathbb{D}|$.

Proof. The left adjoint composed with the evaluation functor at D is given by tensoring with $\mathbb{Z}\mathbb{D}(U_-, D)$. Hence the left adjoint is exact if and only if $\mathbb{Z}\mathbb{D}(U_-, D)$ is flat for all D. By Corollary 2.5 this is true if and only if $\text{aff}(Y, \mathbb{D}(U_-, D))$ has filtered components for all D. But

$$(Y, \mathbb{D}(U_-, D)) = (U, D).$$

Example. To see that some hypothesis is necessary in Theorem 2.2, let $\mathbb{C}$ be the monoid whose elements are 1 and α where $\alpha^2 = \alpha$. Then it is easy to establish an equivalence of categories $\text{Ab}^{\mathbb{C}^{op}} \simeq \text{Ab} \times \text{Ab}$, from which one sees that every right $\mathbb{Z}\mathbb{C}$-module of the form $\mathbb{Z}M$ is projective and hence flat. But if $M = \{x, y, z\}$ where $x\alpha = y\alpha = z\alpha = z$, then (Y, M) is the following category:

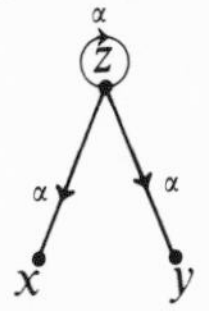

Hence $\text{aff}(Y, M)$ is not filtered in this case.

Remark. If R is any nonzero commutative ring, we can define $\text{aff}_R\,\mathbb{C}$ to be the subcategory of $R\mathbb{C}$ consisting of those morphisms whose coefficients (this time in R) sum to one. Then everything done in this section for $\mathbb{Z}M$ works more generally for RM.

3. Nonadditive Projectivity

If M is a $\mathbb{C}$-set, then a morphism from y to x in the category (Y, M) is identified with a pair (α, x) such that $\alpha x = y$. To avoid this switch in the direction of α, we shall find it more convenient to work with the opposite category $(Y, M)^{op}$.

A *right zero* for a category $\mathbb{C}$ is an endomorphism e of an object that maps to all objects, such that $\alpha e = \alpha' e$ whenever the equation makes sense. Then e is idempotent, and it is easy to see that $\mathbb{C}$ has a right zero if and only if its idempotent completion has an initial object. If M is a $\mathbb{C}$-set, then $(Y, M)^{\mathrm{op}}$ has a right zero if and only if M is a retract of a representable.

A $\mathbb{C}$-set M is *indecomposable* if it is nonempty (that is, if at least one of its values is nonempty), and if it cannot be written as the disjoint union (coproduct) of two nonempty sub $\mathbb{C}$-sets. Any $\mathbb{C}$-set M is the coproduct of its indecomposable sub $\mathbb{C}$-sets, and if M_i are these indecomposables, then (Y, M_i) are the components of (Y, M).

Proposition 3.1. If M is a $\mathbb{C}$-set, then $\mathrm{Hom}_{\mathbb{C}}(M,\ \ )$ preserves epimorphisms if and only if the components of $(Y, M)^{\mathrm{op}}$ have right zeros. Moreover $\mathrm{Hom}_{\mathbb{C}}(M,\ \ )$ preserves finite coproducts if and only if M is indecomposable, in which case $\mathrm{Hom}_{\mathbb{C}}(M,\ \ )$ preserves all coproducts. Hence $\mathrm{Hom}_{\mathbb{C}}(M,\ \ )$ preserves colimits if and only if $(Y, M)^{\mathrm{op}}$ has a right zero, or in other words, if and only if M is a retract of a representable.

Proof. If $\mathrm{Hom}_{\mathbb{C}}(M,\ \ )$ preserves epimorphisms, then the natural map

$$\bigoplus_{x \in M} \mathbb{C}(|x|,\ \ ) \to M$$

splits. The splitting must take each indecomposable subset of M into a single term of the coproduct, from which the first assertion follows. The second assertion is easily established using again the fact that maps from an indecomposable into a disjoint union must each go into one of the terms of the union.

4. Projectivity of $\mathbb{Z}M$

If $\mathbb{M}$ is a monoid, then $\mathbb{Z}$ will be considered as a $\mathbb{Z}\mathbb{M}$-module with elements of $\mathbb{M}$ acting as identities ($\Delta\mathbb{Z}$ in previous notation).

Lemma 4.1 (*Laudal*). If $\mathbb{M}$ is a monoid and $\mathbb{Z}$ is projective, then $\mathbb{M}$ has a right zero.

Proof. The splitting of the augmentation $\mathbb{Z}\mathbb{M} \to \mathbb{Z}$ gives rise to a family of elements $e_1, \ldots, e_k$ of $\mathbb{M}$ and nonzero coefficients $r_1, \ldots, r_k$ in $\mathbb{Z}$ such that $\sum r_i = 1$ (showing $k \geq 1$), and such that

$$\sum r_i \alpha e_i = \sum r_i e_i$$

for all $\alpha \in \mathbb{M}$. Supposing the e_i distinct, we thus see that each $\alpha \in \mathbb{M}$ induces a permutation $\pi(\alpha)$ of the e_i, and π is a monoid homomorphism

$\mathbb{M} \to S(k)$ into the symmetric group on k letters. It is easy to see that $\mathbb{ZN} \to \mathbb{Z}$ splits for every $\mathbb{N}$ that admits a surjective homomorphism (functor) $\mathbb{M} \to \mathbb{N}$. In particular, this applies to the image of π. Now the image of π, being a finite submonoid of a group, is a group. But a group $\mathbb{N}$ for which $\mathbb{ZN} \to \mathbb{Z}$ splits is easily seen to be trivial, and so any of the e_i will serve as a right zero for $\mathbb{M}$.

Theorem 4.2. If M is a $\mathbb{C}$-set, and $M\alpha$ is an injection for all $\alpha \in \mathbb{C}$, then $\mathbb{Z}M$ is projective if and only if the components of $(Y, M)^{\text{op}}$ have right zeros.

Proof. If M is the disjoint union of indecomposables M_i, then $\mathbb{Z}M$ is the coproduct of the $\mathbb{Z}M_i$ and the (Y, M_i) are the components of (Y, M). Hence it suffices to assume that M is indecomposable. Then if $(Y, M)^{\text{op}}$ has a right zero, M is a retract of $\mathbb{C}(C, \ \ )$ for some C, and so $\mathbb{Z}M$ is a retract of $\mathbb{ZC}(C, \ \ )$. Hence $\mathbb{Z}M$ is projective.

Conversely, assume that $\mathbb{Z}M$ is projective where M is indecomposable. Then the natural epimorphism

$$\bigoplus_{x \in M} \mathbb{ZC}(|x|, \ \ ) \xrightarrow{\varepsilon} \mathbb{Z}M$$

splits, and so let μ be a splitting map. Then for each $z \in M$ we have

$$\mu(z) = \sum r_{\alpha, x, z}(\alpha, x), \tag{1}$$

where

$$\sum r_{\alpha, x, z} \alpha x = z, \tag{2}$$

and where if $\beta y = z$, then

$$\sum r_{\alpha, x, y}(\beta\alpha, x) = \sum r_{\alpha, x, z}(\alpha, x). \tag{3}$$

Now under the assumption that the $M\beta$ are injections, we may assume in (1) that the only terms appearing with nonzero coefficients are terms (α, x) with $\alpha x = z$, since if $\beta\alpha x = z = \beta y$, then $\alpha x = y$.

Let us say that x *dominates* z if $\sum_{\alpha x = z} r_{\alpha, x, z} \neq 0$. From (2) we see that every z must be dominated by some element, and from (3) we see that x dominates y if and only if it dominates βy. Then since M is indecomposable, we find that once x dominates one element of M, it dominates every element of M. Let x be such an element. If z is an element of M, let $r_{\gamma, x, z} \neq 0$. Then $\gamma x = z$. If $\beta x = z$ also, then from (3) (with $y = x$) we see that $\beta\alpha = \gamma$ for some α such that $\alpha x = x$. It suffices now to show that the monoid $\mathbb{M} = \{\alpha \mid \alpha x = x\}$ has a right zero e, for then if $\beta x = z$, we have $\beta e = \beta\alpha e = \gamma e$. But since for every $z \in M$ there is a γ with $\gamma x = z$, we obtain a split epimorphism

$$\mathbb{ZC}(|x|, \ \ ) \xrightarrow{\varepsilon} \mathbb{Z}M,$$

and as above, we may assume that the image of x under a splitting map has nonzero coefficients only for those terms α such that $\alpha x = x$. We thus obtain a splitting for $\mathbb{Z}\mathbb{M} \to \mathbb{Z}$, and so by Lemma 4.1 we are done.

The proof of the theorem is modeled after Laudal's proof [5] of the following.

Corollary 4.3. Let $\mathbb{C}$ be any small category. Then lim: $\mathrm{Ab}^{\mathbb{C}} \to \mathrm{Ab}$ is exact if and only if the components of $\mathbb{C}$ have right zeros.

Proof. This follows since lim is given by homing with $\Delta\mathbb{Z} \in \mathrm{Ab}^{\mathbb{C}}$.

Corollary 4.4. If all morphisms of $\mathbb{C}$ are monomorphisms, then $\mathbb{Z}M$ is projective if and only if it is free.

Proof. If all morphisms of $\mathbb{C}$ are monomorphisms and $\mathbb{Z}M$ is projective, then as we saw in Section 2, $M\alpha$ is an injection for all α. Therefore by the theorem, M is a disjoint union of retracts of representables. But again since all morphisms of $\mathbb{C}$ are monomorphisms, the only retracts of representables are the representables themselves. Hence $\mathbb{Z}M$ is free.

Corollary 4.5. If $U: \mathbb{C} \to \mathbb{D}$ takes all morphisms to monomorphisms, then the right adjoint to $\mathrm{Ab}^{\mathbb{D}} \to \mathrm{Ab}^{\mathbb{C}}$ is exact if and only if the components of $(D, U)^{\mathrm{op}}$ have right zeros for all $D \in |\mathbb{D}|$.

Proof. The right adjoint composed with the evaluation functor at D is given by homing with $\mathbb{Z}\mathbb{D}(D, U_{-})$. Hence the right adjoint is exact if and only if the components of $(Y, \mathbb{D}(D, U_{-}))^{\mathrm{op}}$ have right zeros for all D. But $(Y, \mathbb{D}(D, U_{-})) = (D, U)$.

The only place where we required the coefficient ring to be $\mathbb{Z}$ was in Lemma 4.1, where we used the fact that if $\mathbb{N}$ is a group and $\mathbb{Z}$ is projective, then $\mathbb{N}$ is trivial. Here it would suffice to use any ring in which no integer $n > 1$ is invertible. Thus the theorem and its corollaries are valid for such a ring. We shall now consider another version of the theorem which, while putting some hypothesis on the category $\mathbb{C}$, removes all hypothesis from M and the coefficient ring.

Theorem 4.6. Let $\mathbb{C}$ be a category in which the only isomorphisms are identities, and such that an equation $\alpha\beta = \alpha$ implies $\beta = 1$. If R is any nonzero ring and M is a $\mathbb{C}$-set with RM a projective $R\mathbb{C}$-module, then RM is free.

Proof. The hypothesis on $\mathbb{C}$ is easily seen to be equivalent to the single property $\alpha\beta\gamma = \alpha$ implies $\beta = 1$. Let μ be a splitting for the natural epimorphism

$$\bigoplus_{x \in M} R\mathbb{C}(|x|, \ ) \xrightarrow{\varepsilon} RM.$$

Then if y is an element of M, $\mu(y)$ has the form

$$\mu(y) = \sum_{\alpha x = y} r_{\alpha, x, y}(\alpha, x) + \sum_{\alpha x \neq y} r_{\alpha, x, y}(\alpha, x).$$

The sum of the coefficients in the first sum is one and in the second sum is zero. We shall call an element $e \in M$ *minimal* if $\beta z = e$ implies $\beta = 1$ (and hence $z = e$).

Lemma 4.7. If (α, e) appears in the first sum in $\mu(y)$, then e is minimal. Moreover, no term of the form $(1, e)$ with e minimal can appear in the second sum.

Proof. The second statement is clear since there would be nothing to cancel e after application of the augmentation ε. To prove the first assertion, suppose $\beta z = e$. Then $\alpha\beta z = \alpha e = y$, so $\alpha\beta\mu(z) = \mu(y)$. It follows that $\alpha\beta\gamma = \alpha$ for some γ, so $\beta = 1$.

Lemma 4.8. Given a diagram in $(Y, M)^{\text{op}}$

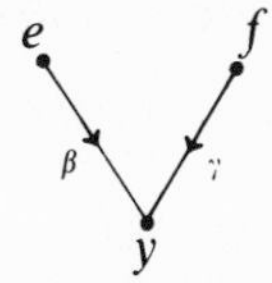

with e and f minimal, we must have $e = f$ and $\beta = \gamma$.

Proof. Since e and f are minimal, $\mu(e)$ and $\mu(f)$ have the form

$$\mu(e) = (1, e) + \sum_{\alpha x \neq e} r_{\alpha, x, e}(\alpha, x)$$

$$\mu(f) = (1, f) + \sum_{\alpha x \neq f} r_{\alpha, x, f}(\alpha, x).$$

Then (β, e) appears in $\mu(y)$, for otherwise $(\beta\alpha, e) = (\beta, e)$ for some (α, e) appearing in the second sum of $\mu(e)$. Then $\alpha = 1$, which is impossible since $(1, e)$ does not appear in the second sum of $\mu(e)$. Thus (β, e) appears in $\mu(y)$, and so $(\beta, e) = (\gamma\alpha, e)$ for some (α, e) appearing in $\mu(f)$. By symmetry, $(\gamma, f) = (\beta\alpha', f)$ for some (α', f) appearing in $\mu(e)$. Thus $\beta = \gamma\alpha = \beta\alpha'\alpha$, so $\alpha' = \alpha = 1$ and $\beta = \gamma$. If $e \neq f$, then since now (β, f) appears in $\mu(y)$, we have $(\beta, f) = (\beta\alpha, f)$ for some (α, f) appearing in $\mu(e)$. But then $\alpha = 1$, contradicting the second statement of Lemma 4.7.

Returning to the proof of the theorem, it follows from Lemmas 4.7 and 4.8 that every object of $(Y, M)^{\text{op}}$ has a unique morphism to it from a unique minimal element of M. It follows easily that each minimal element of M is an initial object for the component that contains it. Thus M is a disjoint union of representables, and so RM is free.

The hypothesis of the theorem is satisfied in the following cases.

1. $\mathbb{C}$ is a *delta*, or in other words, a category whose only endomorphisms and isomorphisms are identities [7]. These include partially ordered sets.
2. $\mathbb{C}$ is the free category generated by a directed graph.
3. $\mathbb{C}$ is the free abelian monoid on any set of generators.

One might conjecture that if K is a field and $\mathbb{C}$ satisfies the hypothesis of the theorem, then all projective $K\mathbb{C}$-modules are free. In Case 1, this is true since $K\mathbb{C}$ is then a local ringoid (that is, each representable has a unique maximal subfunctor). In Case 2 it is true because $K\mathbb{C}$ is a free ideal ringoid (that is, every subfunctor of a representable is a free $K\mathbb{C}$-module). In Case 3, at least when $\mathbb{C}$ is finitely generated, the conjecture is Serre's.

REFERENCES

[1] N. Bourbaki, Eléments de mathématiques; Algébre commutative. *Actualités Sci. Indust.* **1290**, **1293**, Hermann, Paris, 1961.

[2] P. M. Cohn, Free ideal rings. *J. Algebra* **1** (1964), 47–69.

[3] R. Diaconescu, Thesis, Dalhousie University, 1973.

[4] J. Isbell and B. Mitchell, Exact colimits. *Bull. Amer. Math. Soc.* **79** (1973), 994–996.

[5] O. Laudal, Note on the projective limit on small categories. *Proc. Amer. Math. Soc.* **33** (1972), 307–309.

[6] D. Lazard, Autour de la platitude. *Bull. Soc. Math. France* **97** (1969), 81–128.

[7] B. Mitchell, Rings with several objects. *Advances in Math.* **8** (1972), 1–161.

[8] U. Oberst and H. Rohrl, Flat and coherent functors. *J. Algebra* **14** (1970), 91–105.

AMS 18G99

Charles Ching-an Cheng
DEPARTMENT OF MATHEMATICS
OAKLAND UNIVERSITY
ROCHESTER, MICHIGAN

Barry Mitchell
DEPARTMENT OF MATHEMATICS
RUTGERS UNIVERSITY
NEW BRUNSWICK, NEW JERSEY

Some Properties of Two-Dimensional Poincaré Duality Groups

ELDON DYER and A. T. VASQUEZ

In 1967 C. T. C. Wall [17] initiated the formal study of Poincaré complexes. Roughly, these are finite cell-complexes that have Poincaré duality for all constant and locally trivial coefficient groups. In 1972 F. E. A. Johnson and Wall [10] studied Poincaré duality groups, these being groups G for which a $K(G, 1)$ is a Poincaré complex. (See [17] and [10] for precise definitions.) A series of papers by Robert Bieri and by Bieri and Beno Eckmann have extended this study to duality groups.

An interesting case is that of the fundamental group of a closed, two-dimensional surface. Such a surface is a $K(\pi, 1)$ for its fundamental group π, assuming $\pi \neq 1$, and being a closed manifold, it has Poincaré duality for all coefficients. It would be interesting to know whether all two-dimensional Poincaré duality groups were in fact fundamental groups of closed, two-dimensional surfaces. Equivalently, one can ask whether every Poincaré complex with a two-dimensional orientation class has the homotopy type of such a surface.

Let K denote such a complex. J. Cohen has shown [3] that if $H_1(K; \mathbb{Z}) = 0$, then K has the homotopy type of a two sphere, and [4] that if $H_1(K; \mathbb{Z})$ is free abelian of rank two, then K has the homotopy type of a two-dimensional torus. It is also known [5] that if H has index two in G,

G is torsion free, and H is the fundamental group of a closed, two-dimensional surface, then so is G. This reduces the question to that of orientable, two-dimensional Poincaré duality groups.

Specifically then, we consider finitely presented groups G such that there is a class $[G] \in H_2(G; \mathbb{Z})$ for which $\_\_ \cap [G]: H^i(G; \Lambda) \to H_{2-i}(G; \Lambda)$ is an isomorphism for all i and all $\mathbb{Z}[G]$-modules Λ. Rather than repeat the phrase orientable, two-dimensional Poincaré duality groups, we will refer to such groups as *surface groups*.

The object of this paper is to present two results about surface groups.

Theorem A. Let G and G' be surface groups and $a: G \to G'$ be a homomorphism. If $H_*(a; \mathbb{Z}): H_*(G; \mathbb{Z}) \to H_*(G'; \mathbb{Z})$ is an isomorphism, then a is an isomorphism.

Theorem B. For each prime p, every surface group is residually a finite p-group.

In regard to Theorem A we note that it follows easily from Poincaré duality and naturality of cap products that $H_*(a; \mathbb{Z})$ is an isomorphism if $H_1(G; \mathbb{Z})$ and $H_1(G'; \mathbb{Z})$ are isomorphic and $H_2(a; \mathbb{Z})$ is onto. Neither Theorem A nor Theorem B is trivial even for fundamental groups of surfaces: K. Frederick in 1963 [6] proved Theorem B in that case for $p = 2$; G. Baumslag proved the stronger result that for orientable surfaces the fundamental groups are residually free [1]. Theorem A follows from Theorem B by noting that a residually finite p-group is residually nilpotent and applying in this setting a commutator calculus argument (for example see [15]). Actually, Theorem A has a simpler proof than Theorem B and we give that, independently of Theorem B. Also, even for fundamental groups of surfaces the argument for Theorem B is perhaps of interest, resembling somewhat a proof given by Baumslag [2] that a finitely generated group that is a cyclic extension of a free group is residually finite.

1. Preparatory Material

Lemma 1. Let G be an orientable, n-dimensional Poincaré duality group and E a subgroup.

(a) If $[G: E] = \infty$, then $H_n(E; \mathbb{Z}) = 0$.

(b) If $[G: E] = s < \infty$, then the homomorphism $H_n(E; \mathbb{Z}) \to H_n(G; \mathbb{Z})$ can be indentified with $\mathbb{Z} \xrightarrow{xs} \mathbb{Z}$.

Proof. By a Shapiro lemma and Poincaré duality we have

$$H_n(E; \mathbb{Z}) \cong H_n(G; \mathbb{Z}[G] \otimes_{\mathbb{Z}[E]} \mathbb{Z}) \cong H^0(G; \mathbb{Z}[G] \otimes_{\mathbb{Z}[E]} \mathbb{Z}) \cong (\mathbb{Z}[G] \otimes_{\mathbb{Z}[E]} \mathbb{Z})^G.$$

The module $\mathbb{Z}[G] \otimes_{\mathbb{Z}[E]} \mathbb{Z}$ can be indentified with the free abelian group on the cosets G/E with G-module structure arising from the usual action of G on G/E. Thus the module can be identified with integer valued functions on G/E with finite support. Since the action of G on G/E is transitive, the invariant functions are constant. This implies (a).

As the isomorphisms displayed above are natural, we have the commutative diagram

$$\begin{array}{ccc} H_n(E;\mathbb{Z}) & \xrightarrow{\cong} & (\mathbb{Z}[G] \otimes_{\mathbb{Z}[E]} \mathbb{Z})^G \\ \downarrow & & \downarrow \theta \\ H_n(G;\mathbb{Z}) & \xrightarrow{\cong} & (\mathbb{Z}[G] \otimes_{\mathbb{Z}[G]} \mathbb{Z})^G. \end{array}$$

Let $c_1, \ldots, c_s$ denote the distinct cosets of E in G, assuming $[G\colon E] = s < \infty$. Then, as above, we see that $(\mathbb{Z}[G] \otimes_{\mathbb{Z}[E]} \mathbb{Z})^G$ is infinite cyclic and generated by $c_1 + \cdots + c_s$, which is constantly one. By the same token, $(\mathbb{Z}[G] \otimes_{\mathbb{Z}[G]} \mathbb{Z})^G$ is infinite cyclic and generated by c, the unique coset of G in G. Clearly, $\theta(c_1 + \cdots + c_s) = s \cdot c$. ∎

Lemma 2. Let G be a surface group and $\rho\colon G \to \mathbb{Z}$ be an onto homomorphism. Then $K = \ker \rho$ is a free group.

Proof. Let A be a K-module. By another Shapiro lemma and Poincaré duality

$$H^2(K; A) \cong H^2(G; \mathrm{Hom}_{\mathbb{Z}[K]}(\mathbb{Z}[G], A)) \cong H_0(G; \mathrm{Hom}_{\mathbb{Z}[K]}(\mathbb{Z}[G], A)).$$

Let $g \in G$ be such that $\rho g = 1$. As right $\mathbb{Z}[K]$-module,

$$\mathbb{Z}[G] \cong \oplus_n \mathbb{Z}[K]g^n.$$

In order for $w \in \mathrm{Hom}_{\mathbb{Z}[K]}(\mathbb{Z}[G], A)$ to equal $gv - v$ for some $v \in \mathrm{Hom}_{\mathbb{Z}[K]}(\mathbb{Z}[G], A)$, letting $v(g^n) = b_n$ and $w(g^n) = a_n$, it is necessary and sufficient that $a_n = b_{n+1} - b_n$ for all n. Thus, for any w, selecting b_0 arbitrarily, we construct v so that $w = gv - v$. This implies that $H_0(G; \mathrm{Hom}_{\mathbb{Z}[K]}(\mathbb{Z}[G], A)) = 0$ and $H^2(K; A) = 0$ for all K-modules A. By a theorem of R. G. Swan [16] it follows that K is a free group. ∎

It follows routinely by looking at the symplectic form induced by Poincaré duality that for a surface group G, $H_1(G; \mathbb{Z})$ is a free abelian group of even rank, say $2g$, and it is shown in [3] that $g \neq 0$. The number g is called the *genus* of G.

We will indicate the free cyclic group generated by t by $\mathbb{Z}(t)$ and the cyclic group of order k generated by t by $\mathbb{Z}_k(t)$.

Proposition. Let G be a surface group of genus g and

$$1 \to K \to G \xrightarrow{\rho} \mathbb{Z}(t) \to 1$$

be an exact sequence of groups. Then as $\mathbb{Z}_p[\mathbb{Z}(t)]$-module, $H_1(K;\ \mathbb{Z}_p)$ is isomorphic to $F \oplus C$, where F is the direct sum of $2g-2$ free cyclic $\mathbb{Z}_p[\mathbb{Z}(t)]$-modules, and C is a $\mathbb{Z}_p$-vector space of dimension one on which t operates trivially.

Proof. Since G is finitely generated, $H_1(K;\ \mathbb{Z}_p)$ is finitely generated as a module over $\mathbb{Z}_p[\mathbb{Z}(t)]$, which is a principal ideal domain. By the structure theorem for such modules,

$$H_1(K;\ \mathbb{Z}_p) \cong F \oplus C$$

as $\mathbb{Z}_p[\mathbb{Z}(t)]$-modules, where F is a finite direct sum of free cyclic $\mathbb{Z}_p[\mathbb{Z}(t)]$-modules and C is a finite direct sum of torsion modules.

As $\mathbb{Z}_p$-vector space, C is finite dimensional. Thus the automorphism t on C is of some finite order N. Let $\overline{\mathbb{Z}}$ be the subgroup of $\mathbb{Z}(t)$ generated by t^N, $\sigma\colon \overline{\mathbb{Z}} \to \mathbb{Z}(t)$ be the inclusion map, and $\overline{G} = \rho^{-1}\overline{\mathbb{Z}}$. Since $[G\colon \overline{G}] = N < \infty$, it follows by Theorem 2 of [10] that $\overline{G}$ is also a surface group. For $\bar{\rho} = \rho \,|\, \overline{G}\colon \overline{G} \to \overline{\mathbb{Z}}$ and $\overline{K} = \operatorname{Ker} \bar{\rho}$, $K = \overline{K}$ and $H_1(\overline{K};\ \mathbb{Z}_p)$ as module over $\mathbb{Z}_p[\overline{\mathbb{Z}}]$ has action given by t^N. Since

$$\begin{aligned} \mathbb{Z}_p \cong H_2(\overline{G};\ \mathbb{Z}_p) &\cong H_1(\overline{\mathbb{Z}};\ H_1(\overline{K};\ \mathbb{Z}_p)) && \text{(Lyndon spectral sequence)} \\ &\cong H^0(\overline{\mathbb{Z}};\ H_1(\overline{K}\colon \mathbb{Z}_p)) && \text{(Poincaré duality for } \overline{\mathbb{Z}}) \\ &\cong (H_1(\overline{K};\ \mathbb{Z}_p))^{\overline{\mathbb{Z}}} \\ &= C, \end{aligned}$$

we see that C has $\mathbb{Z}_p$-vector space dimension one.

By the same token

$$\mathbb{Z}_p \cong (H_1(K;\ \mathbb{Z}_p))^{\mathbb{Z}} \subset C,$$

since t leaves no nonzero element of F fixed. Thus, $C = (H_1(K;\ \mathbb{Z}_p))^{\mathbb{Z}}$; i.e., t operates trivially on C.

From the Lyndon spectral sequence we have the exact sequence

$$0 \to H_0(\mathbb{Z};\ H_1(K;\ \mathbb{Z}_p)) \to H_1(G;\ \mathbb{Z}_p) \to H_1(\mathbb{Z};\ H_0(K;\ \mathbb{Z}_p)) \to 0.$$

But $H_1(\mathbb{Z};\ H_0(K;\ \mathbb{Z}_p)) \cong \mathbb{Z}_p$, $H_1(G;\ \mathbb{Z}_p)$ is a $\mathbb{Z}_p$-vector space of dimension $2g$, and $H_0(\mathbb{Z}; H_1(K;\ \mathbb{Z}_p))$ is a $\mathbb{Z}_p$-vector space of dimension one greater than the rank of F. Thus F is of rank $2g-2$; i.e., it is the direct sum of $2g-2$ free cyclic $\mathbb{Z}_p[\mathbb{Z}(t)]$-modules. ∎

We note that this proposition implies one of the results of J. Cohen [3]. For if $H_1(G; \mathbb{Z}) \cong \mathbb{Z} \oplus \mathbb{Z}$, then $1 \to K \to G \to \mathbb{Z} \to 1$ is exact with K a free group such that $H_1(K; \mathbb{Z}_p) \cong \mathbb{Z}_p$. This implies $K \cong \mathbb{Z}$. The only extension of $\mathbb{Z}$ by $\mathbb{Z}$ that is a surface group is $\mathbb{Z} \oplus \mathbb{Z}$.

2. Proof of Theorem A

Let $a: G \to G'$ be a homomorphism of surface groups such that $H_*(a; \mathbb{Z})$ is an isomorphism. Then G and G' have the same genus, say g. Let $\rho': G' \to \mathbb{Z}(t)$ be a epimorphism [defined, for example, by abelianization of G' followed by projection onto one of the $2g$ direct summands of $H_1(G'; \mathbb{Z})$] and complete the following commutative diagram as indicated:

$$\begin{array}{ccccccccc} 1 & \to & K & \to & G & \xrightarrow{\rho} & \mathbb{Z}(t) & \to & 1 \\ & & \downarrow{\scriptstyle j} & & \downarrow{\scriptstyle a} & & \downarrow{\scriptstyle =} & & \\ 1 & \to & K' & \to & G' & \xrightarrow{\rho'} & \mathbb{Z}(t) & \to & 1 \end{array}$$

Since $H_2(a; \mathbb{Z})$ is onto, it follows from Lemma 1 that $[G': \text{im } a] = 1$; i.e., a is onto. Hence j is also onto. We will complete the proof by showing j is an isomorphism.

By Lemma 2, K and K' are free groups. Since j is onto, it follows from the theorem of W. Magnus that free groups are residually nilpotent together with Theorem 3.4 of [15] that j is an isomorphism if $H_1(j; \mathbb{Z})$ is an isomorphism. Hence it suffices to show that $H_1(j; \mathbb{Z}_p)$ is an isomorphism for all primes p. (Actually it suffices to show $H_1(j; \mathbb{Z}_p)$ is an isomorphism for one prime p by using the stronger result of Magnus that free groups are residually p-groups and the same theorem of [15].) We know already that $H_1(j; \mathbb{Z}_p)$ is onto.

By the propositon of the previous section we have

$$H_1(K; \mathbb{Z}_p) \cong F_{2g-2} \oplus \mathbb{Z}_p(b) \qquad \text{and} \qquad H_1(K'; \mathbb{Z}_p) \cong F'_{2g-2} \oplus \mathbb{Z}_p(b')$$

as $\mathbb{Z}_p(\mathbb{Z}(t))$-modules, where F_{2g-2} and F'_{2g-2} are each free on $2g-2$ generators and $\mathbb{Z}_p(b)$ and $\mathbb{Z}_p(b')$ are $\mathbb{Z}_p$-vector spaces of dimension one on which t operates trivially.

Let k denote the composition

$$F_{2g-2} \xrightarrow{i} F_{2g-2} \oplus \mathbb{Z}_p(b) \xrightarrow{j_*} F'_{2g-2} \oplus \mathbb{Z}_p(b') \xrightarrow{\pi} F'_{2g-2},$$

where i is inclusion and π projection. Since no nonzero element of F'_{2g-2} is invariant under the action of t and since b is, $\pi j_*(b) = 0$. Since j_* is onto, this implies that k is onto. But k is a module homomorphism of free modules of the same rank over a principal ideal domain. Thus k is an isomorphism.

As noted in the proof of the proposition of the previous section, we have the diagram

$$\begin{array}{ccc} H_2(G; \mathbb{Z}_p) & \xrightarrow{\cong} & (H_1(K; \mathbb{Z}_p))^{\mathbb{Z}} \\ \downarrow \cong & & \downarrow \\ H_2(G'; \mathbb{Z}_p) & \xrightarrow{\cong} & (H_1(K'; \mathbb{Z}_p))^{\mathbb{Z}} \end{array}$$

which is commutative by naturality of the horizontally marked isomorphisms. Thus, $j_* b = ab'$ for some $\alpha \in \mathbb{Z}_p$, $a \neq 0$.

Suppose $j_*(f + \beta b) = 0$ for $f \in F_{2g-2}$ and $\beta \in \mathbb{Z}_p$. Then $0 = \pi j_*(f + \beta b) = \pi j_*(f) = k(f)$. Thus, $f = 0$ and $j_*(\beta b) = 0$. Thus $\beta b = 0$ and Ker $j_* = 0$. ∎

3. Proof of Theorem B

We begin by recalling some definitions and elementary properties from group theory and fixing notation.

For a group G, let $\gamma_1(G) = G$ and for $1 \leq n$, $\gamma_{n+1}(G) = [\gamma_n(G), G] = \{bcb^{-1}c^{-1} \mid b \in \gamma_n(G) \text{ and } c \in G\}$. Let $\gamma_\omega(G) = \bigcap_{n \in \mathbb{Z}^+} \gamma_n(G)$. The sequence $\{\gamma_n(G)\}$, $n \in \mathbb{Z}^+$, is the lower central series of G. For $m \leq n$, $\gamma_n(G)$ is normal in $\gamma_m(G)$; the constructions $\gamma_n(\ )$ and $\gamma_m(\ )/\gamma_n(\ )$ are natural with respect to homomorphisms of groups. The group G is *nilpotent of class n* provided $\gamma_n(G)$ is the trivial group but $\gamma_{n-1}(G)$ is not.

Let Π be a class of groups. The group G is said to be *residually* Π if for each $g \in G$, $g \neq 1$, there is a homomorphism a: $G \to B$ with B in the class Π and $ag \neq 1$. If Π is the class of nilpotent groups, then G is residually Π if and only if $\gamma_\omega(G) = 1$. We shall also be interested in

(1) the class $\mathscr{P}$ of finite p-groups for p a prime and

(2) the class Π_p of groups each element of which has order some power of p.

Let p be a prime and G be a surface group. There is an exact sequence

$$1 \to K \xrightarrow{i} G \xrightarrow{\pi} \mathbb{Z}(t) \to 1$$

with K a free group. Let $g \in G$ and suppose that $g \neq 1$. If $\pi g \neq 1$, then reducing $\mathbb{Z}$ modulo some suitably high power of p we see that g is detected by a homomorphism of G into a finite p-group.

Suppose that $\pi g = 1$; that is, $g \in K$. Since K is residually nilpotent (the previously mentioned theorem of W. Magnus), for some n $g \notin \gamma_n(K)$. The group $K/\gamma_n(K)$ is free in the variety of nilpotent groups of class n. Hence by a theorem of K. Gruenberg [7], $K/\gamma_n(K)$ is residually π_p. Thus for some s, the image $\bar{g}$ of g in the group V, defined to be $K/\gamma_n(K)$ modulo the subgroup of all products of p^s-powers in $K/\gamma_n(K)$, is not one. The group V is free in the variety of nilpotent groups of class n and exponent p^s.

The action of t on K, induced by any lifting of π, defines an action on V and a subsequent action on $V/\gamma_2(V)$. $V/\gamma_2(V)$ is a free $\mathbb{Z}_{p^s}$-module and is also a $\mathbb{Z}_{p^s}[\mathbb{Z}(t)]$-module. Further reduction of $V/\gamma_2(V)$ by its subgroup of pth-powers yields $H_1(K; \mathbb{Z}_p)$; the induced action by t yields the previous $\mathbb{Z}_p[\mathbb{Z}(t)]$-module structure on $H_1(K; \mathbb{Z}_p)$.

By the proposition, $H_1(K; \mathbb{Z}_p) \cong F_{2g-2} \oplus \mathbb{Z}_p(\bar{y})$. Let $\bar{x}_1, \ldots, \bar{x}_{2g-2}$ be a $\mathbb{Z}_p[\mathbb{Z}(t)]$-basis for F_{2g-2}. Then for $\bar{x}_{i,u} \equiv \bar{x}_i t^u$, $u \in \mathbb{Z}$, the set $\{\bar{x}_{i,u}, \bar{y} \mid 1 \le i \le 2g-2, u \in \mathbb{Z}\}$ is a $\mathbb{Z}_p$-basis for $H_1(K; \mathbb{Z}_p)$. Any lifting of this basis to $V/\gamma_2(V)$ is a basis for $V/\gamma_2(V)$ as $\mathbb{Z}_{p^s}$-module. Let y be a lifting of $\bar{y}$ and x_i be a lifting of $\bar{x}_i$, $1 \le i \le 2g-2$. Define $x_{i,u} = x_i t^u$; then $\{x_{i,u}, y \mid 1 \le i \le 2g-2, u \in \mathbb{Z}\}$ is a $\mathbb{Z}_{p^s}$-basis for $V/\gamma_2(V)$. Finally, let Y be a lifting of y and X_i be a lifting of x_i, $1 \le i \le 2g-2$, to V, and define $X_{i,u} = t^{-u} X_i t^u$. Then modulo $\gamma_2(V)$, $\{X_{i,u}, Y \mid 1 \le i \le 2g-2, u \in \mathbb{Z}\}$ is a basis for $V/\gamma_2(V)$. Since V is free in the variety of nilpotent groups of class n and exponent p^s, this implies $\{X_{i,u}, Y \mid 1 \le i \le 2g-2, u \in \mathbb{Z}\}$ freely generated V [9].

Let $-a < 0 < b$ be such that $a + b + 1 = p^r$ for some r and $\bar{g}$, $t(Y)$ and $t^{-1}(Y)$ are all contained in the subgroup W of V generated by $\{X_{i,u}, Y \mid 1 \le i \le 2g-2, -a \le u \le b\}$. Define ρ: $V \to W$ by $Y \mapsto Y$, $X_{i,v} \mapsto X_{i,w}$ for $-a \le w \le b$ and $v = w \bmod p^r$. This is well-defined since the set $\{X_{i,v}, Y\}$ freely generates V. Then ρ is a retraction of V onto W and so $\rho\bar{g} \ne 1$. Again by the freeness of the generators, we can define an endomorphism τ: $W \to W$ by $X_{i,w} \to X_{i,w+1}$ for $-a \le w < b$, $X_{i,b} \mapsto X_{i,-a}$, and $Y \mapsto t(Y)$. Since τ induces an automorphism on $W/\gamma_2(w)$, it is an automorphism, and $\rho t(x) = \tau\rho(x)$ for all $x \in V$ [12, Corollary 42.32].

$H_1(W; \mathbb{Z}_p)$ is a retract of $H_1(V; \mathbb{Z}_p)$ and the induced action of τ on $H_1(W; \mathbb{Z}_p)$ is given by $\bar{x}_{i,w} \mapsto \bar{x}_{i,w+1}$ for $-a \le w < b$, $\bar{x}_{i,b} \mapsto \bar{x}_{i,-a}$, and $\bar{y} \mapsto \bar{y}$. Thus τ^{p^r} is the identity on $H_1(W; \mathbb{Z}_p)$. Since W is a finite p-group, it follows by a theorem of P. Hall [8] that some pth power of τ^{p^r} is the identity on W; i.e., for some positive integer s, τ^{p^s}: $W \to W$ is the identity homomorphism.

The semidirect product $W \cdot Z_{p^s}(\tau)$ is a finite p-group; as it contains W as a subgroup, the image of $\rho\bar{g}$ in $W \cdot \mathbb{Z}_{p^s}(\tau)$ is not one. The composition $G = K \cdot \mathbb{Z}(t) \to V \cdot \mathbb{Z}(t) \to W \cdot \mathbb{Z}_{p^s}(\tau)$ is a homomorphism of G into a finite p-group that carries g to an element different from one. ∎

Corollary. If G is a one-relator surface group, then G is a residually torsion-free nilpotent.

Proof. This is an immediate consequence of Theorem B and results of Labute [11]. ▮

4. The Pro-p-Completion of a Surface Group

For relevant definitions and background the reader is referred to [13] and [14].

Let G be a surface group of genus g and p denote a prime number. Let $\hat{G}$ be the pro-p-completion of G.

Theorem C. $\hat{G}$ is isomorphic to the group generated by generators $x_1, x_2, \ldots, x_{2g-1}, x_{2g}$ subject to one relation $1 = r = [x_1, x_2] \cdot \cdots \cdot [x_{2g-1}, x_{2g}]$. Thus G is isomorphic to the pro-p-completion of the fundamental group of a surface of genus g.

Before giving the proof, we remark that the phrase "generated by ..." is understood in the sense of pro-p-groups. This amounts to the assertion $\hat{G}$ is isomorphic to the quotient of the pro-p-completion of $F(x_1, \ldots, x_{2g})$ by the *closure* of the normal subgroup generated by the word r. That the pro-p-completion of the fundamental group of a surface has the same description follows from the fact that pro-p-completion, being a left adjoint (of the forgetful functor), preserves cokernels. We remark also that the core of the proof is Demůskin's theorem, which classifies what are now called Demůskin groups. We verify simply that $\hat{G}$ is a Demůskin group and that its invariants are $2g$ and p^∞ (see [13]).

Lemma. Let $j\colon G \to \hat{G}$ be the canonical map and let $\hat{G}$ act trivially on $\mathbb{Z}_p$. Then

$$H^q(j)\colon H^q(\hat{G}; \mathbb{Z}_p) \to H^q(G; \mathbb{Z}_p)$$

is an isomorphism for $q = 1$ and 2. Hence $\hat{G}$ is a Demůskin group and its first invariant is $2g$.

Proof. We use the criteria of [14, p. I-15]. Since $x \in H^1(G; \mathbb{Z}_p)$ "is" a homomorphism $f_x\colon G \to \mathbb{Z}_p$, it is trivial that x restricts to zero on $G_0 = \text{kernel}(f_x)$. Thus condition D_1 is true. Hence (condition A_1) $H^1(j)$ is a bijection and $H^2(j)$ an injection. Since $H^2(j)(x \cup y) = H^1(j)(x) \cup H^1(j)(y)$ and the cup product pairing is nondegenerate in $H^*(G; \mathbb{Z}_p)$, it follows that $\text{Im } H^2(j) \neq 0$. But $H^2(G; \mathbb{Z}_p) \cong H_0(G; \mathbb{Z}_p) \cong \mathbb{Z}_p$. Thus $H^2(j)$ is an isomorphism. ▮

Lemma. $\hat{G}$ modulo the closure of the commutator subgroup is isomorphic to the direct sum of $2g$ copies of the p-adic integers. Hence the second invariant of $\hat{G}$ is p^{∞}.

Proof. Since $G/[G, G]$ is isomorphic to the direct sum of $2g$ copies of the integers, this is a trivial consequence of the general isomorphism

$$\widehat{H/[H, H]} \cong \hat{H} \Big/ \overline{[\hat{H}, \hat{H}]}.$$

This isomorphism in turn is a trivial piece of categorical construction. Consider the following commutative diagram of categories and forgetful functors:

$$\begin{array}{ccc} \text{Abelian pro-}p\text{-groups} & \xrightarrow{F_1} & \text{pro-}p\text{-groups} \\ \big\downarrow{\scriptstyle F_2} & & \big\downarrow{\scriptstyle F_3} \\ \text{Abelian groups} & \xrightarrow{F_4} & \text{groups} \end{array}$$

$G \rightsquigarrow \hat{G}$ is left adjoint to F_3 and $H \rightsquigarrow H/\overline{[H, H]}$ is left adjoint to F_1; hence, $G \rightsquigarrow \hat{G}/\overline{[\hat{G}, \hat{G}]}$ is left adjoint to $F_3 \circ F_1$. On the other hand $G \rightsquigarrow G/[G, G]$ is left adjoint to F_4 and $G \rightsquigarrow \hat{G}$ is left adjoint to F_2; hence $G \rightsquigarrow \widehat{G/[G, G]}$ is also left adjoint to $F_3 \circ F_1 = F_4 \circ F_2$. ∎

REFERENCES

[1] G. Baumslag, On generalized free products, *Math. Z.* **78** (1962), 423–438.

[2] G. Baumslag, Finitely generated cyclic extensions of free groups are residually finite, *J. Austral. Math. Soc.* **5** (1971), 87–94.

[3] J. M. Cohen, Poincaré 2-complexes—I, *Topology* **11** (1972), 417–419.

[4] J. M. Cohen, Poincaré 2-complexes—II, preprint.

[5] J. M. Cohen, Homotopy surfaces, preprint.

[6] K. N. Frederick, The Hopfian property for a class of fundamental groups, *Comm. Pure Appl. Math.* **16** (1963), 1–8.

[7] K. W. Gruenberg, Residual properties of infinite soluble groups, *Proc. London. Math. Soc.* **7** (1957), 29–62.

[8] P. Hall, A contribution to the theory of groups of prime-power order, *Proc. London. Math. Soc.* **36** (1933), 29–95.

[9] P. Hall, The splitting properties of relatively free groups, *Proc. London. Math. Soc.* **4** (1954), 343–356.

[10] F. E. A. Johnson and C. T. C. Wall, On groups satisfying Poincaré duality, *Ann. of Math.* **96** (1972), 592–598.

[11] J. P. Labute, On the descending central series of groups with a single defining relation, *J. Algebra.* **14** (1970), 16–23.

[12] H. Neumann, "Varieties of Groups." Springer-Verlag, New York, 1967.

[13] J. -P. Serre, Structure de certains pro-p-groups, *Sem. Bourbaki 1962/63*, n. **252**.
[14] J. -P. Serre, Cohomologie Galoisienne, "Lectures Notes in Math." n. 5, Springer-Verlag, Berlin and New York, 1964.
[15] J. R. Stallings, Homology and central series of groups, *J. Algebra.* **2** (1965), 170–181.
[16] R. G. Swan, Groups of cohomological dimension one, *J. Algebra.* **12** (1969), 585–610.
[17] C. T. C. Wall, Poincaré complexes—I, *Ann. of Math.* **86** (1967), 213–245.

AMS 57B10

DEPARTMENT OF MATHEMATICS
THE GRADUATE SCHOOL
CITY UNIVERSITY OF NEW YORK
NEW YORK, NEW YORK

Properties Invariant within Equivalence Types of Categories

PETER FREYD

All of us know that any "mathematically relevant" property on categories is invariant within equivalence types of categories. Furthermore, we all know that any "mathematically relevant" property on objects and maps is preserved and reflected by equivalence functors. An obvious problem arises: How can we conveniently characterize such properties? The problem is complicated by the fact that the second mentioned piece of common knowledge, that equivalence functors preserve and reflect relevant properties on objects and maps, is just plain wrong.

I first met Sammy in the fall of 1958 and within ten minutes he was selling me on a "stylistic" point that turns out to be the central clue to the problem. (How often Sammy's "stylistic" points have totally changed entire mathematical viewpoints!) It took me 16 years to make the connection.

An equivalence $T\colon \mathbb{A} \to \mathbb{B}$ preserves equalizers but does not reflect them. $T(x)$ can be an equalizer of $T(y)$ and $T(z)$ without x being an equalizer of y and z, albeit for the most perverse of reasons, namely that the sources and targets of x, y, and z do not match as they should in $\mathbb{A}$ (since T can identify objects, they can match in $\mathbb{B}$).

To make the above stated problem amenable, I will restrict attention to elementary sentences in the language of categories, that is, sentences in which all quantifiers refer to objects and maps and the "atomic" predicates are compositions, equality, and source and target assertions. The standard approach to such a problem is to work with the "Frege notation"[1] ($\forall$, $\exists$, $\wedge$, $\vee$, etc.) and attempt an induction (not on sentences, but on formulas in general) on the number of bound variables. We cannot even begin here. Free formulas are not preserved by equivalence functors; in fact, none of the negations of atomic predicates are preserved by equivalence functors.

When I first met Sammy I was working on the metatheorem for abelian categories and he wanted me to state the metatheorem in a certain way. Note that none of us use the Frege notation very much. Note that we do write diagrams on the board and move our arms a bit. Sammy wanted me to formalize the latter. He was right. I must first describe a diagrammatic notation with which to solve the problem. (At the end, as it happens, we can translate back to the Frege notation. But only at the end.)

1. The Diagrammatic Language

By a **graph** I mean a collection of **vertices** together with a collection of **arrows**, each arrow assigned a **source vertex** and a **target vertex**. If one insists upon formalizing this in the standard set-theoretical way, then a graph is a quadruple $\langle V, A, s, t\rangle$, where s and t are functions from A to V.

Any category may be construed as a graph by forgetting compositions. Given a graph G and a category $\mathbb{A}$, a G-**diagram** in $\mathbb{A}$ is a graph homomorphism $D: G \to \mathbb{A}$. We could of course use the free category generated by a graph and turn everything into a discussion of functors. But a finite graph (e.g., one vertex, one arrow) can generate an infinite category (e.g., the monoid of natural numbers), and hence I stick to graphs.

A **path** in a graph is a finite word of arrows $\langle a_1, \ldots, a_n\rangle$ such that the target of a_i is the source of a_{i+1} for $i = 1, \ldots, n-1$. The source of the path is defined as the source of a_1 and the target of the path as the target of a_n. A **commutativity condition** on a graph is an ordered pair of paths each with the same source and target—unless one of the paths is empty, in which case we require that the source and target of the other be equal. A C-**Graph** is a graph together with a set of commutativity conditions. For a C-graph G, a G-diagram in $\mathbb{A}$ is a graph-homomorphism $D: G \to \mathbb{A}$ such that for every commutativity condition $\langle a_1, \ldots, a_n\rangle = \langle b_1, \ldots, b_m\rangle$ it is the case that $D(a_1)D(a_2) \cdots D(a_n) = D(b_1) \cdots D(b_m)$—unless $m = 0$, in which case we require that $D(a_1) \cdots D(a_n)$ be an identity map.

[1] Frege's notation, of course, was very different. The phrase "Frege notation," however, has come into standard use.

The following diagrams appear in Fig. 1:

$$xy = xz$$

$$xy = xz$$
$$ay = az$$

$$xy = xz$$
$$ay = az$$
$$bx = a$$

$$xy = xz$$
$$ay = az$$
$$bx = a$$
$$b'x = a$$

$$xy = xz$$
$$ay = az$$
$$bx = a$$
$$b'x = a$$
$$b = b'$$

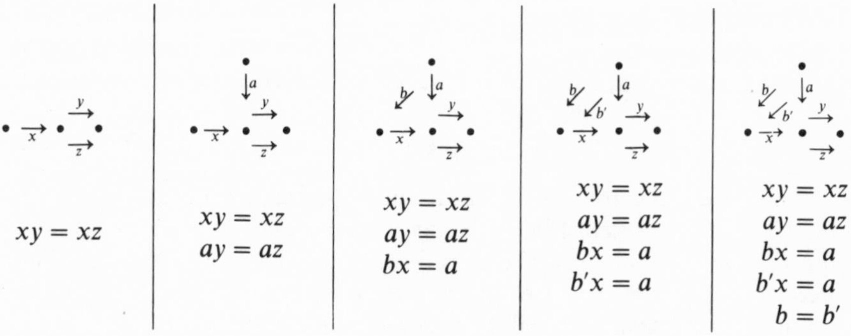

Fig. 1

Consider the nested family of C-graphs shown in Fig. 1. A diagram from the first is an equalizer diagram if and only if for every extension to the second there exists an extension to the third such that for every extension to the fourth there is an extension to the fifth.

A standard simplification of notation is to assume that every conceivable commutativity condition holds unless we say otherwise. I will say otherwise by inserting question marks within the graph, where it is to be understood that the question mark removes only one commutativity condition, namely that which immediately surrounds it. Figure 2 has three commutativity

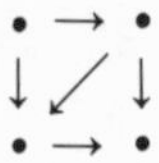

Fig. 2   **Fig. 3**   **Fig. 4**

conditions, Fig. 3 has two, and Fig. 4 still has one (the outer square). We may define a product diagram using Fig. 5. That is, a diagram from the first graph is a product if and only if for all extensions to the second

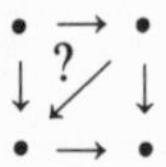

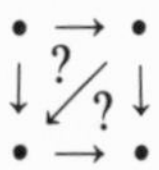

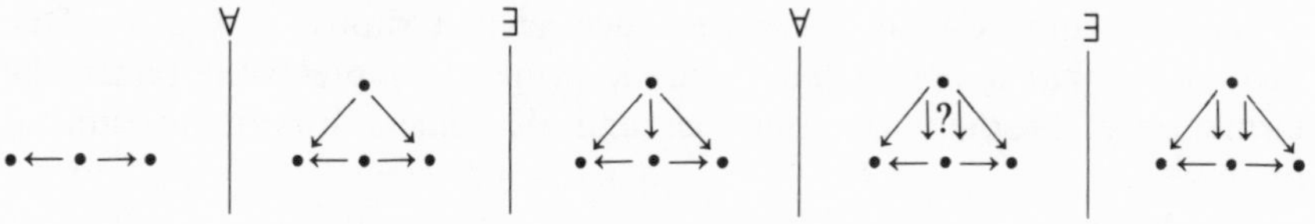

Fig. 5

there is an extension to the third such that for all extensions to the fourth there is an extension to the fifth. The last two C-diagrams are needed for the uniqueness, and when we arm-wave at a blackboard we customarily omit them and say the word "unique." Hence Fig. 6 defines product diagrams.

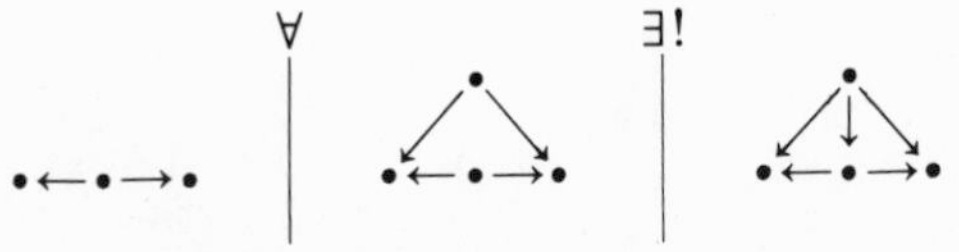

Fig. 6

A finite *rooted tree*, recall, is a finite partially ordered set with a unique smallest element (the *root*) such that two elements have a common upper bound if and only if one is less than the other. The immediate successors of the root will be called the *near-roots*; the tree that sprouts upwards from a near-root will be called its *corresponding subtree*.

A *CG*-**tree** is a finite rooted tree of *C*-graphs ordered by extension, each labeled by ∀ or ∃. We define the notion that a diagram $D: R \to \mathbb{A}$, where R is the root, **satisfies** the tree, recursively, as follows:

If the root is labeled ∃(∀), $D: R \to \mathbb{A}$ *satisfies the tree if an* (*if every*) *extension of D to a near-root satisfies the corresponding sub-CG-tree.*

If the tree is just its root, then $D: R \to \mathbb{A}$ satisfies the tree if and only if the root is labeled ∀.

If R is empty then the tree describes a property on categories, namely that the empty diagram satisfies the tree. For example, the linear tree of Fig. 7 is satisfied by $\varnothing \to \mathbb{A}$ if and only if $\mathbb{A}$ has binary products. We will call such properties **diagrammatic properties**.

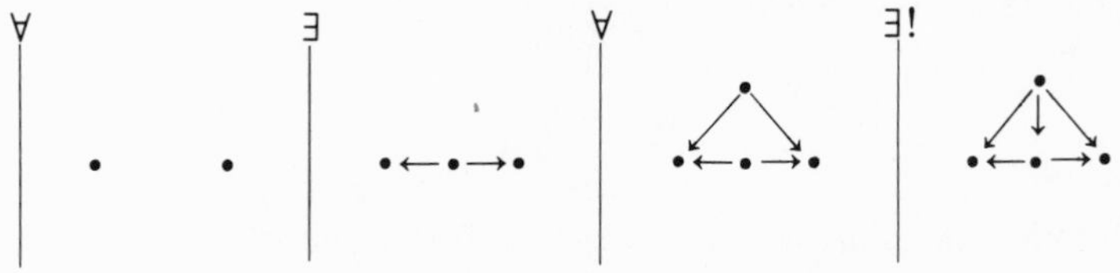

Fig. 7

Linear trees do not suffice. For example, the property that $\mathbb{A}$ is linearly connected requires a nonlinear tree such as that shown in Fig. 8. (One may check that a linear diagrammatic property is preserved under the formation of products of categories and that linear ordering is not so preserved.)

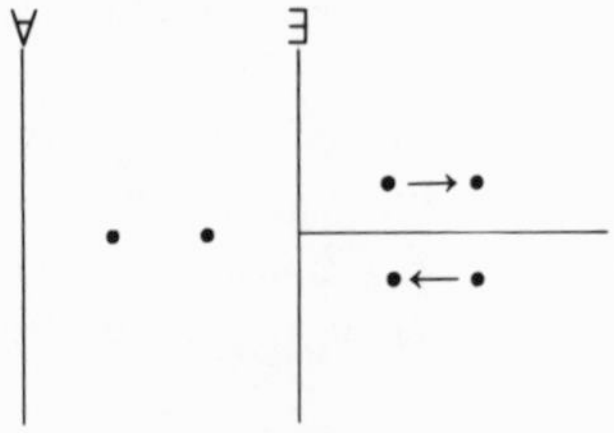

Fig. 8

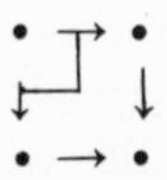

Fig. 9

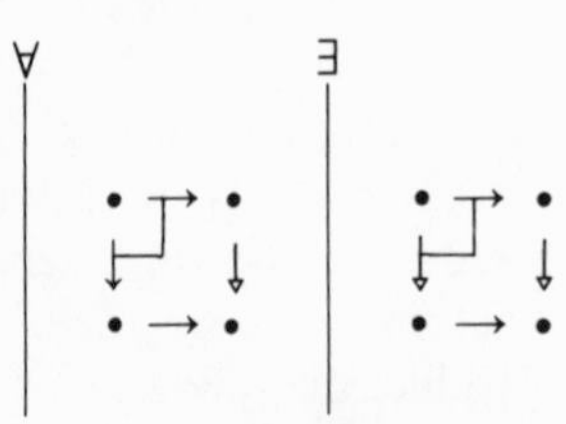

Fig. 10

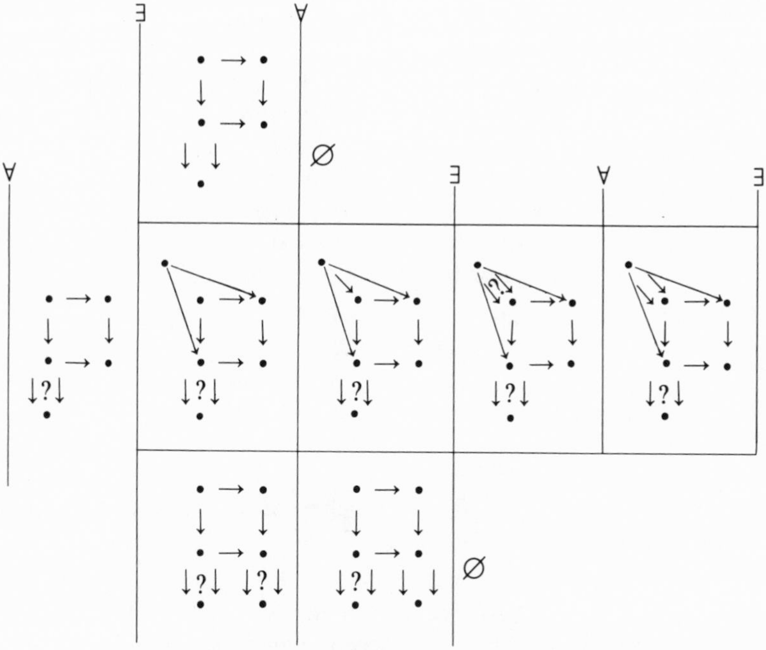

Fig. 11

Given a CG-tree T with root R define the complementary tree T' as that obtained by transposing $\forall$ and $\exists$. Then $D\colon R \to \mathbb{A}$ satisfies T' if and only if it does *not* satisfy T. Diagrammatic sentences are closed under the usual Boolean operators of negation, conjunction, and disjunction. If one excepts the source-target information in the root, then the Boolean operators are available for CG-trees. Note that the labels $\forall$ and $\exists$ serve both as quantifiers and as conjunctions and disjunctions. Over the years we have developed notations to avoid nonlinear trees. For example, if Fig. 9 denotes a pullback and $\twoheadrightarrow$ an epimorphism, then the property that pullbacks transfer epimorphisms is that shown in Fig. 10. If one does not use such notation, then we are forced to the nonlinear tree in Fig. 11.

2. The Theorem

Theorem. An elementary property on categories is invariant within equivalence types of categories if and only if it is a diagrammatic property.

Outline of Proof. Induction does now work for the easy direction. That is, if $F\colon \mathbb{A} \to \mathbb{B}$ is an equivalence of categories (use only that it is full, faithful with a representative image) for any CG-tree T with root R and diagram $D\colon R \to \mathbb{A}$ that satisfies T, then $R \to A \xrightarrow{F} \mathbb{B}$ also satisfies T. The diagrammatic notation successfully avoids the anomalies that result from the Frege notation.

For the other direction, we define a CI-graph as a C-graph together with a distinguished set of arrows, called "identity conditions." If G is a CI-graph, then $D: G \to \mathbb{A}$ is a diagram if besides respecting the commutativity conditions it carries the distinguished arrows into identity maps in $\mathbb{A}$. Just as above we define CIG-trees and what it means for a diagram from the root to satisfy a CIG-tree. Say that a graph is simple if each vertex appears as a source or target at most once.

Lemma. For any elementary property $P(A_1, \ldots, A_n, x_1, \ldots, x_m)$ there is a CIG-tree T with simple root R with $\langle a_1 \cdots a_m \rangle$ as arrows, $\{v_1, \ldots, v_n, sa_1, ta_1, \ldots, sa_n, ta_n\}$ as vertices, such that $D: R \to \mathbb{A}$ satisfies T if and only if

$$P(D(v_1), \ldots, D(v_n), D(a_1), \ldots, D(a_m))$$

is true in $\mathbb{A}$. In particular, for every elementary sentence S there is a CIG-tree T with empty root such that $\varnothing \to \mathbb{A}$ satisfies T if and only if $\mathbb{A}$ satisfies S.

Lemma. For every CIG-tree T with empty root there is a CG-tree T' with empty root such that for all skeletal categories $\mathbb{A}$, $\varnothing \to \mathbb{A}$ satisfies T if and only if it satisfies T'.

This is the difficult lemma. One proves by a cumbersome induction over all trees, empty-rooted or not, that for every CIG-tree T with root R there is a CIG-tree T' with root R such that D: $R \to \mathbb{A}$ satisfies T if and only if it satisfies T' for all skeletal $\mathbb{A}$, where T' is such that all identity conditions involve only arrows that appear in the root. Hence if R is empty then T' is a CG-tree. T' tends to be much fatter than T.

The lemmas yield the theorem: If S is a sentence invariant within equivalence types, let T be an empty-rooted CG-tree such that $\varnothing \to \mathbb{A}$ satisfies T if and only if $\mathbb{A}$ satisfies S for all *skeletal* $\mathbb{A}$. Since every category is equivalent to a skeletal category and $\varnothing \to \mathbb{A}$ satisfying T is invariant within equivalence types and, by assumption, so is S, then $\varnothing \to \mathbb{A}$ satisfies T if and only if $\mathbb{A}$ satisfies S for all $\mathbb{A}$, skeletal or not. (By using the Gödel completeness theorem one needs only that all *countable* categories are equivalent to skeletal categories, and hence can avoid using the axiom of choice.)

3. Back to Frege

Consider the Frege language on two sorts: "objects," A, B, C, ...; "maps," x, y, z, ...; and atomic predicates $(x = y)$, $(A = B)$, $(xy = z)$, $(A = \square x)$, $(A = x\square)$, where the last two are pronounced "A is the source (target) of x." We wish to characterize those sentences invariant within equivalence type.

We shall interpret the "restricted quantifiers,"

$$\forall_{A \xrightarrow{x} B}[\cdots] \quad \text{as} \quad \forall_x[(A = \Box x) \wedge (B = x\Box) \Rightarrow \cdots]$$

and

$$\exists_{A \xrightarrow{x} B}[\cdots] \quad \text{as} \quad \exists_x[(A = \Box x) \wedge (B = x\Box) \wedge \cdots].$$

Note that $\neg \forall_{A \xrightarrow{x} B}[\cdots]$ is equivalent with $\exists_{A \xrightarrow{x} B} \neg [\cdots]$. A sentence will be called a **Frege-diagrammatic** sentence if all quantified maps are so restricted and

(1) No map is quantified without its source and target having been previously quantified;

(2) The atomic predicates $(A = \Box x)$, $(B = \Box x)$ do not appear other than implicitly in the restricted quantifiers;

(3) The atomic predicate $(A = B)$ does not appear;

(4) If $(x = y)$ appears as an atomic predicate then the restricted quantifiers for x and y imply that $\Box x = \Box y$ and $x\Box = y\Box$;

(5) If $(xy = z)$ appears as an atomic predicate then the restricted quantifiers for x, y, and z imply that $\Box x = \Box z$, $x\Box = \Box y$, and $y\Box = z\Box$;

(6) If $x = 1_A$ appears as an atomic predicate then the restricted quantifier implies $A = \Box x$ and $A = x\Box$.

It is routine that for an empty-rooted CG-tree T there is a Frege-diagrammatic sentence S such that $\varnothing \to \mathbb{A}$ satisfies T if and only if $\mathbb{A}$ satisfies S. Conversely, we can find for any Frege-diagrammatic sentence such a CG-tree. Hence, an elementary sentence S is invariant within equivalence types if and only if there is a Frege-diagrammatic sentence S' such that the axioms of category theory imply $S \Leftrightarrow S'$. There can be no algorithm, incidentally, for deciding whether an arbitrary sentence is invariant within equivalence types. (For any word problem for monoids there is a sentence S true for all categories if and only if the given word problem is true. $S \vee \forall_{A, B}(A = B)$ is invariant within equivalence types if and only if S is true for all categories.)

Linear CG-trees correspond to prenex Frege-diagrammatic sentences, that is, all quantities in front. The sentence

$$\forall_{A, B}[(\exists_{A \xrightarrow{x} B}(x = x)) \vee (\exists_{B \xrightarrow{x} A}(x = x))]$$

says that a category is linearly connected. It cannot be put in prenex Frege-diagrammatic form.

AMS 02-16, 18-10

DEPARTMENT OF MATHEMATICS
UNIVERSITY OF PENNSYLVANIA
PHILADELPHIA, PENNSYLVANIA

Coherence for the Tensor Product of 2-Categories, and Braid Groups

JOHN W. GRAY

Introduction

In [4], a tensor product is defined between 2-categories that is part of a (nonsymmetric) monoidal closed category structure on the category of (small) 2-categories and 2-functors (i.e., categories and functors enriched in the category of small categories.) The only point that is not fully treated in [4] is the coherence of the associativity of this tensor product. The purpose of this paper is to provide a complete proof of this coherence, ab initio, without use of [7], by means of a faithful representation of a certain category in the positive semigroup of an appropriate braid group. The result follows by a solution of the word problem for the image of this representation. In a subsequent paper it will be shown that this also implies a result about the structure of certain 2-theories, which in turn implies a number of standard coherence theorems. I would like to thank several colleagues who have helped to educate me in the subtleties of word problems, especially K. Appel, who provided the proof of 3.1.2, and J. Rotman, who told me about [3].

In Section 1, the construction of the tensor product is outlined and the precise conditions for coherence of its associativity are stated. In

Section 2, a 2-category $\mathbf{Q}_N$ called the N-dimensional cube is constructed, and it is shown that the precise conditions amount to $\mathbf{Q}_N$ being commutative, i.e., locally partially ordered. This is reduced to showing that a suitable representation in a braid group is faithful. In Section 3 this is proven by introducing a class of admissable words and showing that they can be reduced to canonical form, while also showing that there is a section of the projection onto the symmetric group whose image is exactly the set of canonical forms.

1. The Tensor Product

In [4], quasi-functors of n-variables are introduced together with their representing tensor product. These are defined as follows.

1.1. A quasi-functor of two variables $H\colon \mathbf{A} \times \mathbf{B} \to \mathbf{C}$, between 2-categories, consists of families of 2-functors

$$\{H(A, \_\_)\colon \mathbf{B} \to \mathbf{C} \mid \text{for all } A \in \mathbf{A}\}$$
$$\{H(\_\_, B)\colon \mathbf{A} \to \mathbf{C} \mid \text{for all } B \in \mathbf{B}\}$$

such that $H(A, \_\_)(B) = H(\_\_, B)(A) =_{\text{def}} H(A,B)$, for all A and B, together with 2-cells

$$\gamma_{f,g}\colon H(f, B')H(A, g) \to H(A', g)H(f, B),$$

for all $f\colon A \to A'$, $g\colon B \to B'$, satisfying

(a) $\gamma_{1,g} = 1$, $\gamma_{f,1} = 1$ (where 1 denotes an appropriate identity map),

(b) $\gamma_{f'f,g} = \gamma_{f',g}H(f, B) \cdot H(f', B')\gamma_{f,g} =_{\text{def}} \gamma_{f',g} \boxminus \gamma_{f,g}$
$\gamma_{f,g'g} = H(A', g')\gamma_{f,g} \cdot \gamma_{f,g'} H(A, g) =_{\text{def}} \gamma_{f,g'} \mathbin{\square\!\!\!\!\mid\,} \gamma_{f,g}$

(c) if $u\colon f \to f'$ and $v\colon g \to g'$ are 2-cells, then

$$H(A', g')H(\mu, B) \cdot H(A', v)H(f, B) \cdot \gamma_{f,g}$$
$$= \gamma_{f',g'} \cdot H(f', B')H(A, v) \cdot H(\mu, B')H(A, g).$$

Here, juxtaposition denotes composition in $\mathbf{C}$, while $(\cdot)$ denotes composition within the hom categories of $\mathbf{C}$. Juxtaposition is carried out first. See [4] for diagrams representing these conditions.

1.2. A quasi-functor of n-variables, $n > 2$, $H\colon \prod_{i=1}^n \mathbf{A}_i \to \mathbf{C}$, consists of quasi-functors of two variables

$$H(A_1, \ldots, A_{i-1}, \_\_, A_{i+1}, \ldots, A_{j-1}, \_\_, A_{j+1}, \ldots, A_n)\colon \mathbf{A}_i \times \mathbf{A}_j \to \mathbf{C}$$

for all $i < j$ and all choices of indicated objects $A_k \in \mathbf{A}_k$, which agree on objects and as 2-functors of one variable, and such that for all triples of indices $i < j < k$ and all morphisms

$$f_i\colon A_i \to A_i', \qquad f_j\colon A_j \to A_j', \qquad f_k\colon A_k \to A_k'$$

one has (omitting extraneous variables)

$$\begin{aligned}\gamma_{A_i'f_jf_k} H(f_i, A_j, A_k) \cdot H(A_i', f_j, A_k')\gamma_{f_iA_jf_k} \cdot \gamma_{f_if_jA_k'} H(A_i, A_j, f_k)\\ = H(A_i', A_j', f_k)\gamma_{f_if_jA_k} \cdot \gamma_{f_iA_j'f_k} H(A_i, f_j, A_k) \cdot H(f_i, A_j', A_k')\gamma_{A_if_jf_k}.\end{aligned}$$

This says that certain cubes with 2-cells in their faces commute; namely, cubes of the form shown in Fig. 1. Here, all possible abbreviations are used;

$$H(i, j, k) = H(A_i, A_j, A_k), \qquad H(f_i, 1, 1) = H(f_i, A_j, A_k),$$

$$\gamma_{i,j,1} = \gamma_{f_i, f_j, A_k}, \qquad \text{etc.}$$

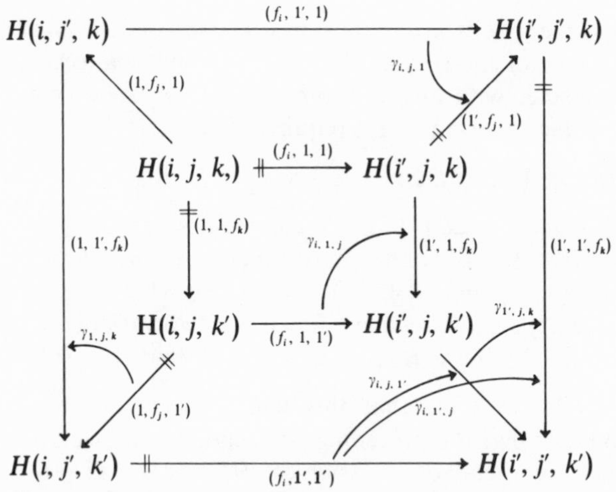

Fig. 1

The notched arrows indicate the two composed 1-cells that are the common domain and codomain of the two 2-cells whose equality is required.

It is easily checked that if $F_i\colon \prod_{j=1}^{n_i} \mathbf{A}_{ij} \to \mathbf{B}_i$ is a quasi-functor of n_i-variables and $G\colon \prod_{i=1}^{n} \mathbf{B}_i \to \mathbf{C}$ is a quasi-functor of n-variables, then

$$G(F_1, \ldots, F_n)\colon \prod_{i=1}^{n} \left(\prod_{j=1}^{n_i} \mathbf{A}_{ij}\right) \to \mathbf{C}$$

is a quasi-functor of $\left(\sum_{i=1}^{n} n_i\right)$-variables. Hence these operations form a "good" system of "multilinear functions."

1.3. The tensor product $\mathbf{A} \otimes \mathbf{B}$ of a pair of 2-categories is the 2-category such that

(a) Objects are ordered pairs (A, B), $A \in \mathbf{A}$, $B \in \mathbf{B}$;

(b) Morphisms are equivalence classes of composable words in ordered pairs (f, g) where f and g are morphisms in $\mathbf{A}$ and $\mathbf{B}$, respectively, and either f or g is an identity morphism. Words are equivalent if they are made so by the smallest equivalence relation compatible with juxtaposition such that $(f, 1)(f', 1) \sim (ff', 1)$ and $(1, g)(1, g') \sim (1, gg')$;

(c) 2-cells are generated by ordered pairs (σ, τ) where σ and τ are 2-cells in $\mathbf{A}$ and $\mathbf{B}$, respectively, with either σ or τ an identity 2-cell, together with 2-cells

$$\gamma_{f,g}\colon (f, B')(A, g) \to (A', g)(f, B)$$

for all pairs of morphisms f and g in $\mathbf{A}$ and $\mathbf{B}$, respectively.

Formally, the 2-cells are equivalence classes of strings of equivalence classes of strings of these symbols subject to the relations satisfied by a 2-category together with one additional type of relation corresponding to 1.1(c) that (in the notation there) requires that

$$(A', g')(\mu, B) \cdot (A', v)(f, B) \cdot \gamma_{f,g} \sim \gamma_{f',g'} \cdot (f', B')(A, v) \cdot (\mu, B')(A, g)$$

holds for all μ and v. (See [4] for a complete description.)

If $J\colon \mathbf{A} \times \mathbf{B} \to \mathbf{A} \otimes \mathbf{B}$ is the quasi-functor of two variables given by $J(\_\_, B)(\sigma) = (\sigma, B)$ and $J(A, \_\_)(\tau) = (A, \tau)$, where $\gamma_{f,g}$ is the unique so-labeled 2-cell, then, for any quasi-functor of two variables $H\colon \mathbf{A} \times \mathbf{B} \to \mathbf{C}$, there is a unique 2-functor $\bar{H}\colon \mathbf{A} \otimes \mathbf{B} \to \mathbf{C}$ such that $H = \bar{H}J$.

1.4. In order to show that this tensor product is associative, it is sufficient to show that the preceding correspondence extends to bijections between quasi-functors of 3-variables $H\colon \mathbf{A} \times \mathbf{B} \times \mathbf{C} \to \mathbf{D}$ and quasi-functors of 2-variables $\bar{H}\colon (\mathbf{A} \otimes B) \times \mathbf{C} \to \mathbf{D}$ and $\bar{H}'\colon \mathbf{A} \times (\mathbf{B} \otimes \mathbf{C}) \to \mathbf{D}$, respectively. We treat only the first case: given H, set $\bar{H}((A, B), C) = H(A, B, C)$ and on generators for 1-cells and 2-cells.

$\bar{H}((f, g), h) = H(f, g, h)$ where f or $g = 1$, and if f or $g \neq 1$ then $h = 1$.

$\bar{H}((\sigma, \tau), \mu) = H(\sigma, \tau, \mu)$ where σ or $\tau = 1$, and if σ or $\tau \neq 1$, then $\mu = 1$.

$\bar{H}(\gamma_{f,g}, C) = \gamma_{f,g,C}$, $C \in \mathbf{C}$.

The structural 2-cells $\bar{\gamma}$ for $\bar{H}$ are defined by

$$\bar{\gamma}_{(f,1),h} = \gamma_{f,1,h}, \qquad \bar{\gamma}_{(1,g),h} = \gamma_{1,g,h}.$$

The only point where it is not obvious that this defines a quasi-functor of two variables is the compatibility condition 1.1(c) for 2-cells of the form $\mu = \gamma_{f,g}$, $\nu = 1_h$; i.e., for the 2-cell

$$(\gamma_{f,g}, h): ((f, B')(A, g), h) \to ((A', g)(f, B), h)$$

the equation

$$\bar{H}((A', B'), h)\bar{H}(\gamma_{f,g}, \mathrm{C}) \cdot 1 \cdot [\bar{\gamma}_{(A',g),h} \boxplus \bar{\gamma}_{(f,B),h}]$$
$$\overset{?}{=} [\bar{\gamma}_{(f,B'),h} \boxplus \bar{\gamma}_{(A,g),h}] \cdot \bar{H}(\gamma_{f,g}, C)\bar{H}((A, B), h)$$

must be satisfied. But using the definitions, this equation translates precisely into the commutative cube condition in 1.2. Therefore there is an associativity isomorphism

$$\alpha: \mathbf{A} \otimes (\mathbf{B} \otimes \mathbf{C}) \to (\mathbf{A} \otimes \mathbf{B}) \otimes \mathbf{C}.$$

1.5. Finally, it must be shown that this isomorphism is coherent; i.e., all diagrams constructed by successive applications of instances of α commute. It is well-known from [7] that this holds if and only if a certain pentagon commutes, and it is immediate that this holds if and only if the above bijections extend to bijections between quasi-functors of four variables $K: \mathbf{A} \times \mathbf{B} \times \mathbf{C} \times \mathbf{D} \to \mathbf{E}$ and quasi-functors of three variables

$$K': (\mathbf{A} \otimes \mathbf{B}) \times \mathbf{C} \times \mathbf{D} \to \mathbf{E}$$
$$K'': \mathbf{A} \times (\mathbf{B} \otimes \mathbf{C}) \times \mathbf{D} \to \mathbf{E}$$
$$K''': \mathbf{A} \times \mathbf{B} \times (\mathbf{C} \otimes \mathbf{D}) \to \mathbf{E},$$

respectively.

Given K, K' is constructed in exactly the same way $\bar{H}$ was constructed from H in 1.4, except that there is one more variable. To show that K' is a quasi-functor of three variables, it must be shown that cubes, as in 1.2, commute. The only question arises when one of the maps is of the form, say, $(f, B')(A, g)$ in $\mathbf{A} \otimes \mathbf{B}$. The appropriate 2-cells comparing this with $h: C \to C'$ and $k: D \to D'$ are

$$\gamma_{f,B',h,D} \boxplus \gamma_{A,g,h,D} \qquad \text{and} \qquad \gamma_{f,B',C,k} \boxplus \gamma_{A,g,C,k},$$

respectively. It can be shown, using the identities satisfied in a 2-category, that the required equation holds. However, the calculation is long and there are many other similar equations to be checked. It turns out that it is conceptually simpler to imagine a 4-dimensional cube analogous to the 3-dimensional one in 1.2, constructed by including one more variable, $k: D \to D'$, with all the appropriate 2-cells. This hypercube has 3-dimensional faces that are all instances of the commutative cubes in 1.2, and what is required is that certain composed 2-cells in the 4-dimensional cube be the same. Thus the desired result is a consequence of the following statement.

1.6. Proposition. A 4-dimensional cube, all of whose 3-dimensional faces commute, is commutative.

Actually, we could dispense with all reference to other coherence results if we could show that quasi-functors of n-variables correspond to quasi-functions of $(n-1)$-variables by forming tensor products of two adjacent factors and arguing as above. What is needed then is the following generalization, which is proven in the remainder of the paper.

1.7. Proposition. An n-dimensional cube, $n \geq 3$, all of whose 3-dimensional faces commute, is commutative.

2. The N-Dimensional Cube $\mathbf{Q}_N$ as a 2-Category

2.1. Definition. The *free N-dimensional cube with commutative 3-dimensional faces*, $\mathbf{Q}_N$, is the 2-category described as follows:

(a) Objects of $\mathbf{Q}_N$ are sequences

$$J = (i_1, \ldots, i_N), \qquad i_k = 0, 1$$

0 denotes the sequence with all 0s and **1** the sequence with all 1s. If J has a zero in the mth place, then $J(m)$ is the sequence agreeing with J except in the mth place where it has a 1. In general, if J has zeros in places m_i, $i = 1, \ldots, k$, then inductively

$$J(m_k, \ldots, m_1) = J(m_{k-1}, \ldots, m_1)(m_k).$$

Note that the order of the m_i is irrelevant.

(b) Morphisms (or 1-cells) of $\mathbf{Q}_N$ are freely generated by basic 1-cells

$$t_{m,J}: J \to J(m), \qquad m = 1, \ldots, N$$

for each sequence J with an 0 in the mth place; i.e., morphisms are (associative) words in these basic 1-cells.

(c) 2-cells of $\mathbf{Q}_N$ are given as follows: if J has zeroes in the mth and nth places, $m < n$, then there is a basic 2-cell $t_{m,n,J}$ as indicated,

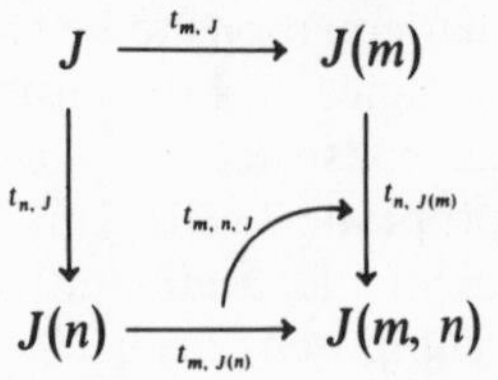

i.e., $t_{m,n,J}: t_{m,J(n)} t_{n,J} \to t_{n,J(m)} t_{m,J}$.

The 2-cells of $\mathbf{Q}_N$ are the required compositions (strings of strings) of these basic 2-cells with each other and with 1-cells, subject to the axioms of a 2-category, together with the relations for all $m < n < p$,

$$\left(t_{n,\,p,\,J(m)}\,t_{m,\,J}\right)\cdot\left(t_{n,\,J(m,\,p)}\,t_{m,\,p,\,J}\right)\cdot\left(t_{m,\,n,\,J(p)}\,t_{p,\,J}\right)$$
$$=\left(t_{p,\,J(m,\,n)}\,t_{m,\,n,\,J}\right)\cdot\left(t_{m,\,p,\,J(n)}\,t_{n,\,J}\right)\cdot\left(t_{m,\,J(n,\,p)}\,t_{n,\,p,\,J}\right).$$

(This says that all 3-dimensional cubes commute.)

2.2. Theorem. $\mathbf{Q}_N$ is locally partially ordered (i.e., the hom-categories of $\mathbf{Q}_N$ are partially ordered categories).

Remark. This clearly implies Proposition 1.6 since, given any N-dimensional cube in a 2-category $\mathbf{X}$ with commutative 3-dimensional faces, there is a 2-functor from $\mathbf{Q}_N$ to $\mathbf{X}$. Since $\mathbf{Q}_N$ commutes (i.e., there is at most one 2-cell between any pair of 1-cells) so does the given cube.

Proof of 2.2. The proof comes in several steps whose details are treated in Section 3.

2.3. Step 1. By induction on N, it is sufficient to treat the category $\mathbf{C} = \mathbf{Q}_N(\mathbf{0}, \mathbf{1})$. Our first task is to give a better description of this category. The objects of $\mathbf{C}$ are the 1-cells of $\mathbf{Q}_N$ from $\mathbf{0}$ to $\mathbf{1}$; i.e., composable words in basic 1-cells $t_{m,\,J}$. Now, each $t_{m,\,J}$ turns a 0 into a 1, and there are no 1-cells going the opposite way. Hence a 1-cell from $\mathbf{0}$ to $\mathbf{1}$ can be represented by a permutation

$$A = a_1, \ldots, a_N$$

of $I = (1, \ldots, N)$, where a_m stands for the basic 1-cell

$$t_{a_m,\,\mathbf{0}(a_{m+1},\,\ldots,\,a_N)}.$$

For example, if $N = 3$,

$$231 \equiv t_{2,\,\mathbf{0}(3,\,1)}\,t_{3,\,\mathbf{0}(1)}\,t_{1,\,\mathbf{0}}.$$

Intuitively, the A are interpreted as the various ways of successively switching N switches from 0 to 1, so $\mathbf{C}$ has $N!$ objects.

To encode the morphisms of $\mathbf{C}$ (the 2-cells of $\mathbf{Q}_N$ between the above 1-cells) observe that a basic 2-cell $t_{m,\,n,\,J}$ gives rise to such a 2-cell by preceding J by a path of basic 1-cells from $\mathbf{0}$ to J and following $J(m, n)$ by a path of basic 1-cells from it to $\mathbf{1}$. In terms of the above representation, if A and B are permutations that agree except at two successive places, say m and $m+1$, and $a_m < a_{m+1}$, $a_m = b_{m+1}$, $a_{m+1} = b_m$, then there is a 2-cell from A to B,

$$\sigma_{mAB} = t_{a,\,K} \cdots t_{a_m a_{m+1}\,J(a_{m+2})} t_{a_{m+2}\,J} \cdots t_{a_N\mathbf{0}},$$

where $J = \mathbf{0}(a_{m+3}, \ldots, a_N)$ and $K = \mathbf{0}(a_2, \ldots, a_N)$. Here m ranges from 1 to $N-1$. The morphisms of $\mathbf{C}$ are the composable words in the σ_{mAB}, subject to two types of relations.

R1. If $|m-n| = 1$ and $\sigma_{mCD}\sigma_{nBC}\sigma_{mAB}$ is defined, then there exist unique B' and C' with $\sigma_{nC'D}\sigma_{mB'C'}\sigma_{nAB'}$ defined and these are equal. (This is the relation for 3-dimensional cubes.)

R2. If $|m-n| \geq 2$ and $\sigma_{mBC}\sigma_{nAB}$ is defined, then there is a unique B' with $\sigma_{nB'C}\sigma_{mAB'}$ defined and these are equal. (This is because $\mathbf{Q}_N$ is a 2-category.)

2.4. Step 2. Let B_N denote the braid group on N strings, [1], [2], and [3]; i.e., B_N is generated by σ_m, $m = 1, \ldots, (N-1)$, with the relations

R'1. $\sigma_m\sigma_{m+1}\sigma_m = \sigma_{m+1}\sigma_m\sigma_{m+1}, \quad m = 1, \ldots, N-2$

R'2. $\sigma_m\sigma_n = \sigma_n\sigma_m \quad$ if $|m-n| \geq 2$.

Regard B_N as a category with a single object and let $P\colon \mathbf{C} \to B_N$ be the functor such that $P(\sigma_{mAB}) = \sigma_m$ for all A, B, and m, P is well-defined since whenever two morphisms are equal in $\mathbf{C}$ as a consequence of *R1* and *R2*, their images are equal in B_N as a consequence of *R'1* and *R'2*. In Section 3, the following will be shown.

2.5. Lemma. $P\colon \mathbf{C} \to B_N$ is faithful.

This means that if $f, g\colon A \to B$ in $\mathbf{C}$ satisfy $P(f) = P(g)$, then $f = g$. It will be shown that $P(f) = P(g)$ implies that there is a sequence of words $w_1, \ldots, w_n$ in the generators of B_N such that $w_1 = P(f)$, $w_n = P(g)$ and w_{i+1} is obtained from w_i by applying either *R'1* or *R'2*. By induction, *R1* and *R2* imply that $w_i = P(f_i)$ for some f_i, and that

$$f = f_1 = f_2 = \cdots = f_n = g$$

holds in $\mathbf{C}$; i.e., applications of *R'1* and *R'2* in B_N can be lifted to $\mathbf{C}$ as applications of *R1* and *R2*. The difficulty is that, in general, equality in B_N means equality modulo the normalizer of the subgroup generated by the relations, and there is no way to lift insertions of terms of the form $\sigma\sigma^{-1}$.

2.6. Step 3. Let S_N denote the symmetric group on N letters. We regard S_N as the quotient of B_N by the projection $P'\colon B_N \to S_N$ given by adding the relation

R'3. $\sigma_m{}^2 = 1, \quad m = 1, \ldots, (N-1)$

to those defining B_N. See [3]. The following will be shown in Section 3.

2.7. Lemma. (i) The composition $P'P: \mathbf{C} \to S_N$ is faithful.

(ii) If $f: A \to B$ is a morphism in $\mathbf{C}$, then $P'P(f)$ is the permutation taking A to B.

The proof of Theorem 2.2 is now immediate. Since if f, $g: A \to B$ in $\mathbf{C}$, then $P'P(f) = P'P(g)$ and hence $f = g$, so $\mathbf{C}$ is partially ordered.

3. Proofs of the Lemmata

The method of proof is to construct canonical forms for elements in the image of P. Note that the image of P consists of positive words in the generators (no σ_m^{-1} occur), so we must work within the positive semigroup of B_N. Hence the elegant solution of the word problem for B_N by Artin [1] and Bohnenblust [2] does not apply to our situation. The proof is completed by showing that the canonical form for a morphism from A to B depends only on the permutation taking A to B.

3.1. Admissable Words. Let F_N denote the free group on generators $\sigma_1, \ldots, \sigma_{N-1}$, and $Q: F_N \to B_N$ the projection given by imposing the relations *R'1* and *R'2* of 2.4. Let X_N denote the set of all permutations of $I = (1, \ldots, N)$ and let F_N operate on X_N (on the left) by taking σ_m to the permutation $(m, m+1)$ regarded as an operation on any $A \in X_N$. If $A = (a_1, \ldots, a_N) \in X_N$ and $S \in F_N$, then the jth term of $S(A)$ is denoted by $S(A)_j$.

3.1.1. Definition. An element $S \in F_N$ is called *A-admissible* if

(i) S is positive; i.e., all exponents of σ_m in S are positive.
(ii) For all decompositions $S = S_1 \sigma_j S_2$,

$$S_2(A)_j < S_2(A)_{j+1};$$

i.e., at each stage S moves a smaller integer to the right of a larger one.

S is called *admissible* if it is I-admissible. $\Sigma(A)$ denotes the set of A-admissible words and Σ the set of admissible ones.

3.1.2. Proposition. $\Sigma(A) \subset \Sigma$.

Proof. We shall show that S is A-admissible if and only if A satisfies a finite set of conditions of the form

$$C(i, j) = (i < j \text{ and } a_i < a_j),$$

depending on S. Since I satisfies all of the conditions $C(i, j)$, the result follows.

To prove this, suppose $S = \sigma_{i_1} \cdots \sigma_{i_p}$ is A-admissible. Then

(a) σ_{i_p} is A-admissible so A satisfies $C(i_p, i_p + 1)$.

(b) Write $S = S_1 \sigma_j S_2$ where $j = i_k$. Then σ_j is $S_2(A)$-admissible, so $S_2(A)_j < S_2(A)_{j+1}$. Now $S_2(A)_j = a_{m_j}$ and $S_2(A)_{j+1} = a_{n_j}$ for some m_j and n_j, so the A-admissibility of S is equivalent to $a_{m_j} < a_{n_j}$ for all j. What must be shown is that $m_j < n_j$, so that this is condition $C(m_j, n_j)$. Suppose $m_j > n_j$. If so then, since in $S_2(A)$ the image of a_{m_j} preceds a_{n_j}, an interchange must have taken place in S_2 moving the larger a_{n_j} to the right of the smaller a_{m_j}, contradicting the A-admissibility of S.

3.1.3. Proposition. (i) In an admissible word all exponents are $+1$.

(ii) A subword of an admissible word is admissible.

(iii) If S is admissible and S' results from S by applying one of the relations *R′1* and *R′2* in 2.4, then S' is admissible.

(iv) If σ_i occurs twice in an admissible word, then between any two consecutive occurrences of σ_i, there is a σ_{i+1} or a σ_{i-1}.

Proof. (i) Clear, since $\sigma_i \sigma_i$ is obviously not A-admissible for any A.

(ii) If $S = S_1 S_2 S_3$ is admissible, then S_2 is $S_3(I)$-admissible, and hence, by 3.1.2, admissible.

(iii) It is easily seen that $\sigma_m \sigma_{m+1} \sigma_m$ and $\sigma_{m+1} \sigma_m \sigma_{m+1}$ are both A-admissible if and only if $a_m < a_{m+1} < a_{m+2}$. Similarly, $\sigma_m \sigma_n$ and $\sigma_n \sigma_m$, with $|m - n| \geq 2$, are both A-admissible if and only if $a_m < a_{m+1}$ and $a_n < a_{n+1}$.

(iv) If there is no σ_{i+1} or σ_{i-1}, then *R′2* can be used to bring the two σ_i into juxtaposition, contradicting (i).

3.2. Canonical Forms.

3.2.1. Definition. (cf. [2].) (i) A word $S \in F_N$ is called *k-pure* if it is of the form

$$S = S_{jk} = \sigma_j \sigma_{j+1} \cdots \sigma_k, \qquad j \leq k.$$

(This differs from [3].)

(ii) A word $S \in F_N$ is in *normal form* if it is of the form

$$S = S_{k_r} \cdots S_{k_1},$$

where

(a) S_{k_i} is k_i-pure;

(b) $k_r > \cdots > k_2 > k_1$.

3.2.2. Proposition. If S is in normal form, then S is admissible.

Proof. A k-pure word S_{jk} acts on I by moving $k + 1$ to the left until it is in the jth position. Hence it is admissible. If S is in normal form, then S

acts by first moving the smallest integer $k_1 + 1$ to the left, then moving $k_2 + 1 > k_1 + 1$ to the left, etc. This is clearly admissible.

3.2.3. Proposition. If S is admissible, then there is a sequence of admissible words in F_N, $S_1, \ldots, S_n$, such that

(a) $S_1 = S$;
(b) S_{i+1} results from S_i by applying either $R'1$ or $R'2$ (see 2.4);
(c) S_n is in normal form.

In particular, in B_N, $S = S_1 = \cdots = S_n$.

Proof. The proof is by induction on the length (number of σ_i) and the number of occurrences of the σ_i with maximum index (value of i). For an admissible word of length 1, there is nothing to prove. Let S be an admissible word of length m and assume the result for all admissible words of shorter length. Let M be the maximum index of a σ_i in S.

Case 1. There is only one occurrence of σ_M.

(i) If σ_M is first, reduce the rest of the word [which is admissible by 3.1.3(ii)] to normal form. The entire word then is in normal form.

(ii) If σ_M is not first, put the initial (left-hand) segment into normal form. Then either

(a) σ_M can be moved to the beginning using $R'2$, which reduces to (i), or
(b) σ_M moves to the left using $R'2$ to an $(M-1)$-pure word. This followed by σ_M is M-pure. Then treat the rest of the word by induction.

Case 2. There are two or more occurrences of σ_M. By 3.1.3(iv), between any two occurrences of σ_M there is a σ_{M-1}.

(i) Treat the first σ_M as in Case 1.

(ii) Reduce the word between the first and second σ_M to normal form. This begins with an $(M-1)$-pure segment, all other segments being of lower index. Hence the second σ_M can be moved to the left using $R'2$ to obtain a segment

$$\sigma_M \sigma_j \cdots \sigma_{M-1} \sigma_M, \qquad j \leq M-1.$$

Use $R'2$ to write this as

$$\sigma_j \cdots \sigma_M \sigma_{M-1} \sigma_M$$

and then $R'1$ to get

$$\sigma_j \cdots \sigma_{M-1} \sigma_M \sigma_{M-1}.$$

This has one less σ_M than the original word, which gives the result by induction.

Next, we examine the relation between normal forms and permutations and show that every permutation is represented by a unique normal form.

3.2.4. Proposition. $P'Q\colon F_N \to S_N$ has a section $\Gamma\colon S_N \to F_N$ whose image consists of the set of normal forms.

Proof. Let $\sigma \in S_N$ be the permutation taking $I = (1, \ldots, N)$ to B. $\Gamma(\gamma)$ is constructed as follows.

(i) Denote I by A_1. Let n_1 be the smallest integer that is preceded by an $m < n_1$ in A_1 but not in B. Let m_1 be the smallest such m. In A_1, $n_1 = a_{k_1}$, $m_1 = a_{j_1}$ with $j_1 < k_1$. Then the first (right-hand) segment of $\Gamma(\sigma)$ is

$$\sigma_{j_1} \cdots \sigma_{k_1 - 1}.$$

Denote the result of applying this to A_1 by A_2.

(ii) Given A_r, let n_r be the smallest integer that is preceded by an $m < n_r$ in A_r but not in B. Let m_r be the leftmost such m. In A_r, $n_r = a_{k_r}$, $m_r = a_{j_r}$ with $j_r < k_r$. Then the rth segment from the right in $\Gamma(\sigma)$ is

$$\sigma_{j_r} \cdots \sigma_{k_r - 1}.$$

This process terminates with B in at most $N - 1$ steps. For example, if $N = 5$ and $B = 25341$ then, divided into segments,

$$\Gamma(\sigma) = \sigma_2 \sigma_3 \sigma_4 | \sigma_3 | \sigma_2 | \sigma_1.$$

Clearly $P'Q\Gamma(\sigma) = \sigma$; i.e., as permutations, $\Gamma(\sigma)$ and σ agree. Furthermore, by construction, the segments of $\Gamma(\sigma)$ are pure, and since we began with the integers in natural order and $\sigma_{j_r} \cdots \sigma_{k_r - 1}$ affects only the first k_r entries, we have

$$k_r > \cdots > k_2 > k_1.$$

Therefore, $\Gamma(\sigma)$ is in normal form. Finally, it is clear that if S is in normal form, then $\Gamma P'Q(S) = S$.

3.3. Conclusion of the Proof. Consider the diagram

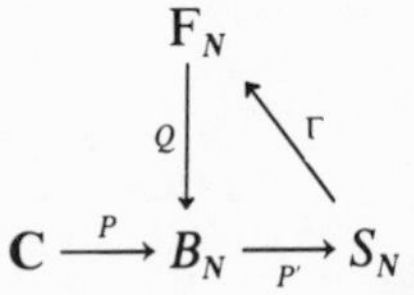

(i) By construction of the 2-cells σ_{mAB} in 2.3, it follows, in the present terminology, that when σ_{mAB}: $A \to B$, then $B = \sigma_m(A)$. Hence in general, for any f: $A \to B$ in $\mathbf{C}$, $P'P(f)$ is the permutation taking A to B. Hence for all $f, g \in \mathbf{C}(A, B)$, $P'P(f) = P'P(q)$.

(ii) Also by construction, σ_{mAB} only acts to move a smaller entry to the right of a larger one, so $P(\mathbf{C}) \subset Q(\Sigma)$. It follows from 3.2.3 and the discussion following 2.5 that any f: $A \to B$ in $\mathbf{C}$ is equal to an f_n: $A \to B$, where $P(f_n) = Q(S_f)$ and S_f is a uniquely determined normal form. Call f_n the normal form of f.

(iii) Since $S_f = \Gamma P'Q(S_f)$, it follows that if $f, g \in \mathbf{C}(A, B)$, then

$$\begin{aligned} P(f) &= P(f_n) = Q(S_f) = Q\Gamma P'Q(S_f) = Q\Gamma P'P(f_n) \\ &= Q\Gamma P'P(g_n) = Q(S_g) = P(g_n) = P(g). \end{aligned}$$

(iv) Finally we shall show that if $P(f_n) = P(g_n)$ then $f_n = g_n$ and hence $f = g$. For $P(f_n) = P(g_n)$ implies $Q(S_f) = Q(S_g)$ as above, and hence that $S_f = S_g$, since Q restricted to normal forms is injective ($P'Q = \Gamma^{-1}$). But S_f determines f_n uniquely, since if

$$S_f = \sigma_{i_p} \cdots \sigma_{i_1},$$

then

$$f_n = \sigma_{i_p A_{i_{p-1}} A_{i_p}} \cdots \sigma_{i_2 A_1 A_2} \sigma_{i_1 A A_1},$$

where

$$A_j = \sigma_{i_j} \cdots \sigma_{i_1}(A).$$

This concludes the proof.

Remark. It is easy to describe explicitly when $\mathbf{C}(A, B)$ is nonempty; namely, there is a map from A to B if and only if for all m and n with $m < n$, if n precedes m in A then n precedes m in B. Furthermore, when $\mathbf{C}(A, B) \neq \varnothing$ then any two words in basic 2-cells leading from A to B have the same length, since they can be obtained from each other using $R1$ and $R2$, neither of which changes the length of words. This says that if a permutation B can be obtained from A by interchanges moving a smaller number to the right of a larger one, then all sequences of interchanges accomplishing this have the same length.

REFERENCES

[1] E. Artin, Theory of braids, *Ann. of Math.* **48** (1947), 101–126.

[2] F. Bohnenblust, The algebraic braid group, *Ann. of Math.* **48** (1947), p. 127–136.

[3] H. S. M. Coxeter and W. O. J. Moser, "Generators and Relations for Discrete Groups," 3rd ed. Springer-Verlag, New York, 1972.

[4] J. W. Gray, Formal category theory—Adjointness for 2-categories, "Springer Lecture Notes in Mathematics." Vol. 391 (1974). Springer-Verlag, New York.
[5] J. W. Gray, 2-Algebraic theories and triples, *Cahiers Topologie Geom. Différentielle* **XIV** (1974), 178–180.
[6] H. Langmaack, Verbandstheoretische Einbettung von Klassen unwesentlick verschiedener Ableitungen in die Zopfgruppe, *Computing* **7** (1971), 293–310.
[7] S. Mac Lane, Natural associativity and commutativity, *Rice Univ. Studies* **49** (1963), 28–46.

This work has been partially supported by NSF Grant GP-33143.

AMS 18D05

DEPARTMENT OF MATHEMATICS
UNIVERSITY OF ILLINOIS AT URBANA-CHAMPAIGN
URBANA, ILLINOIS

Homology of Certain H-Spaces as Group Ring Objects

DALE HUSEMOLLER

Dedicated to Sammy Eilenberg on the occasion of his 60th birthday celebration.

This paper originated from conversations with Graeme Segal while I was trying to understand Quillen's work on the Adams conjecture and the calculation of the algebraic K-theory of finite fields. The notion of ring objects in the category of commutative coalgebras, also called Hopf rings, was first studied by J. Milgram [4] in relation to properties of $H_*(BG)$ and $H_*(G)$ for $G = \Omega^\infty S^\infty$.

Here we will show rather explicitly the existence of a "group ring" object for connected groups in the category of commutative coalgebras, i.e., cocommutative Hopf algebras. This is a special case of a more general theory of group rings in a category that we will take up at a later time. Our interest here is to give some idea of their possible importance for the study of the homology of certain H-spaces.

The remainder of the paper is devoted to the study of examples of group rings. Besides the rather elementary example of $H_*(\mathbf{Z} \times BU)$, we will see that Quillen's calculations of the K-theory $K_*(k)$ of a finite field k can be further understood by using this notion of group ring.

1. Ring Objects

Let $\mathscr{X}$ denote a category with finite products, or equivalently, a category with a final object such that the product $X \sqcap Y$ exists in $\mathscr{X}$ for any two objects X and Y. Let $\mathrm{gr}(\mathscr{X})$ and $\mathrm{ab}(\mathscr{X})$ denote respectively the categories of groups and abelian groups in $\mathscr{X}$, see [3, pp. 3, 4] for definitions. For example, if $\mathscr{X} =$ (sets) the category of sets, then gr(sets) = (gr) the category of groups, and ab(sets) = (ab) the category of abelian groups.

Let k denote throughout a commutative ring with 1, which in most considerations will be a field. We denote by (cc) the category of positively graded commutative coalgebras over k, and $(\mathrm{cc})_0$ the full subcategory of connected coalgebras. The final object is k and the unique morphism $\varepsilon(C)\colon C \to k$ is the unit of the coalgebra, and the product $C \sqcap D$ of two coalgebras is $C \otimes D$ with projections

$$C = C \otimes k \xleftarrow{C \otimes \varepsilon(D)} C \otimes D \xrightarrow{\varepsilon(C) \otimes D} k \otimes D = D.$$

If C and D are connected, then so is $C \otimes D$. The coproduct of a family $(C_i)_{i \in I}$ of commutative coalgebras is formed from the coproduct $\coprod_{i \in I} C_i$ of the underlying graded k-modules, but in the connected case it is formed from $k \oplus \coprod_{i \in I} I(C_i)$ where $I(C_i) = \ker(\varepsilon(C_i)\colon C_i \to k)$ the part of C_i in positive degrees.

Now gr(cc) and ab(cc) are just the categories of cocommutative and bicommutative Hopf algebras over k, respectively. In the definition of Hopf algebra we include the antipodal morphism that exists automatically in the connected case, see [5]. This says that every monoid in $(\mathrm{cc})_0$ is a group.

Definition 1. A ring object R in a category $\mathscr{X}$ with finite products is a triple $(R, \varphi_a(R), \varphi_m(R))$ where the following hold:

(A) The pair $(R, \varphi_a(R))$ is a commutative group object in $\mathscr{X}$, that is, it is in $\mathrm{ab}(\mathscr{X})$.

(M) The operation $\varphi_m(R)\colon R \sqcap R \to R$ is associative, that is, the following diagram is commutative:

$$\begin{array}{ccc} R \sqcap R \sqcap R & \xrightarrow{\varphi_m(R) \sqcap R} & R \sqcap R \\ {\scriptstyle R \sqcap \varphi_m(R)} \downarrow & & \downarrow {\scriptstyle \varphi_m(R)} \\ R \sqcap R & \xrightarrow{\varphi_m(R)} & R \end{array}$$

(AM) The distributative laws hold, that is, the following diagrams are commutative:

$$
\begin{array}{ccc}
R \Pi (R \Pi R) & \xrightarrow{R \Pi \varphi_a(R)} & R \Pi R \\
\downarrow{\scriptstyle \Delta \Pi (R \Pi R)} & & \\
(R \Pi R) \Pi (R \Pi R) & & \downarrow{\scriptstyle \varphi_m(R)} \\
\downarrow{\scriptstyle R \Pi T \Pi R} & & \\
R \Pi R \Pi R \Pi R & & \\
\downarrow{\scriptstyle \varphi_m(R) \Pi \varphi_m(R)} & & \\
R \Pi R & \xrightarrow{\varphi_a(R)} & R
\end{array}
\qquad
\begin{array}{ccc}
(R \Pi R) \Pi R & \xrightarrow{\varphi_a(R) \Pi R} & R \Pi R \\
\downarrow{\scriptstyle (R \Pi R) \Pi \Delta} & & \\
(R \Pi R) \Pi (R \Pi R) & & \downarrow{\scriptstyle \varphi_m(R)} \\
\downarrow{\scriptstyle R \Pi T \Pi R} & & \\
R \Pi R \Pi R \Pi R & & \\
\downarrow{\scriptstyle \varphi_m(R) \Pi \varphi_m(R)} & & \\
R \Pi R & \xrightarrow{\varphi_a(R)} & R
\end{array}
$$

Let k denote the final object of $\mathscr{X}$, anticipating our main example. As usual the zero morphism $\eta_a(R)\colon k \to R$ and the inverse morphism $R \to R$ of $\varphi_a(R)$ are uniquely determined by $\varphi_a(R)$. A ring object R has a unit $\eta_m(R)\colon k \to R$ provided it is a unit for $\varphi_m(R)\colon R \Pi R \to R$. If the unit exists, then it is unique, and we will speak of *rings with unit*. A ring object R is *commutative* provided $\varphi_m(R)T = \varphi_m(R)$ where $T\colon R \Pi R \to R \Pi R$ interchanges the factors.

Definition 2. A morphism $f\colon R \to R'$ of ring objects in $\mathscr{X}$ is a morphism $f\colon R \to R'$ in $\mathscr{X}$ such that the following two diagrams are commutative:

$$
\begin{array}{ccc}
R \Pi R & \xrightarrow{f \Pi f} & R' \Pi R' \\
\downarrow{\scriptstyle \varphi_a(R)} & & \downarrow{\scriptstyle \varphi_a(R')} \\
R & \xrightarrow{f} & R'
\end{array}
\qquad
\begin{array}{ccc}
R \Pi R & \xrightarrow{f \Pi f} & R' \Pi R' \\
\downarrow{\scriptstyle \varphi_m(R)} & & \downarrow{\scriptstyle \varphi_m(R')} \\
R & \xrightarrow{f} & R'
\end{array}
$$

When R and R' are rings with unit, we say that $f\colon R \to R'$ is a morphism of rings with unit if in addition $f\eta_m(R) = \eta_m(R')$.

With respect to $f\eta_m(R) = \eta_m(R')$ recall that it is automatic that $f\eta_a(R) = \eta_a(R')$ and that f commutes with the corresponding inverse operations to φ_a.

If $f\colon R \to R'$ and $g\colon R' \to R''$ are two morphisms of rings or rings with unit, then so is $gf\colon R \to R''$ defined in $\mathscr{X}$. Thus with Definitions 1 and 2 we are led to $\mathrm{rg}(\mathscr{X})$ the *category of rings* (*ring objects*) in $\mathscr{X}$ and $\mathrm{rg}_1(\mathscr{X})$ the *category of rings with unit* in $\mathscr{X}$. These categories in turn each have a full subcategory determined by the commutative rings denoted respectively:

$$\mathrm{crg}(\mathscr{X}) \subset \mathrm{rg}(\mathscr{X}) \qquad \text{and} \qquad \mathrm{crg}_1(\mathscr{X}) \subset \mathrm{rg}_1(\mathscr{X}).$$

Note that $\mathrm{rg}_1(\mathscr{X})$ is a subcategory of $\mathrm{rg}(\mathscr{X})$ that is *not* full except in trivial cases.

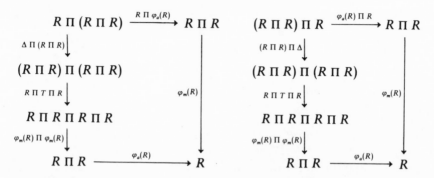

Remark. Let $\mathscr{X}$ and $\mathscr{Y}$ be categories with finite products, and let $F: \mathscr{X} \to \mathscr{Y}$ be a functor preserving finite products. Then F induces functors also denoted simply by F:

$$\begin{array}{ll} \mathrm{gr}(\mathscr{X}) \to \mathrm{gr}(\mathscr{Y}) & \mathrm{ab}(\mathscr{X}) \to \mathrm{ab}(\mathscr{Y}) \\ \mathrm{rg}(\mathscr{X}) \to \mathrm{rg}(\mathscr{Y}) & \mathrm{crg}(\mathscr{X}) \to \mathrm{crg}(\mathscr{Y}) \\ \mathrm{rg}_1(\mathscr{X}) \to \mathrm{rg}_1(\mathscr{Y}) & \mathrm{crg}_1(\mathscr{X}) \to \mathrm{crg}_1(\mathscr{Y}). \end{array}$$

For example, if $R = (R, \varphi_a(R), \varphi_m(R))$ is a ring in $\mathscr{X}$, then $F(R) = (F(R), F(\varphi_a(R))\theta^{-1}, F(\varphi_m(R))\theta^{-1})$ where $\theta: F(R \sqcap R) \to F(R) \sqcap F(R)$ is the natural morphism that is an isomorphism when F preserves finite products, and so $F(R)$ is a ring in $\mathscr{Y}$. If $\eta_m(R): k \to R$ is a unit for R, then $F(\eta_m(R)): F(k) \to F(R)$ is a unit for $F(R)$ since $F(k)$ is a final object in $\mathscr{Y}$ and the required relation holds.

Examples. For $\mathscr{X}$ = (sets) the category of sets, rg(sets) = (rg), crg(sets) = (crg), rg_1(sets) = (rg_1), and crg_1(sets) = (crg_1) the usual categories of rings. The category of sets is related to the category (cc) of commutative coalgebras over k by a product preserving functor $k[\ ]$: (sets) $\to$ (cc). Namely $k[X]$ is the k-module with basis X and the structure morphisms

$$\Delta: k[X] \to k[X] \otimes k[X] \qquad \text{and} \qquad \varepsilon: k[X] \to k$$

are defined by the relations $\Delta(x) = x \otimes x$ and $\varepsilon(x) = 1$. The coalgebras $k[X]$ are called *constant coalgebras* and their elements are concentrated in degree 0. Clearly the natural morphism $k[X \times Y] \to k[X] \otimes k[Y]$ is an isomorphism that says that $k[\ ]$ is product preserving. By the above remark k induces functors

$$(\mathrm{gr}) \to \mathrm{gr}(\mathrm{cc}), \ldots, (\mathrm{crg}_1) \to \mathrm{crg}_1(\mathrm{cc}).$$

This provides us with a family of examples of rings in (cc).

Of particular importance is the ring $\mathbf{Z}$ of integers since it is the initial object in the categories (rg_1) and (crg_1). Its image $k[\mathbf{Z}]$ in crg_1(cc) is denoted simply by $k(\varepsilon)$ and its natural k-basis is denoted by $(\varepsilon^n)_{n \in \mathbf{Z}}$. Moreover

$$\Delta(\varepsilon^n) = \varepsilon^n \otimes \varepsilon^n, \qquad \varphi_a(\varepsilon^i \otimes \varepsilon^j) = \varepsilon^{i+j}, \qquad \text{and} \qquad \varphi_m(\varepsilon^i \otimes \varepsilon^j) = \varepsilon^{ij}.$$

As an object in ab(cc), the Hopf algebra $k(\varepsilon)$ is also the quotient of the polynomial k-algebra $k[x, x^{-1}] = k[x, y]/(xy - 1)$. For any ring object R with unit in (cc) we have a unique morphism $\eta(R): k \to R$ where $\eta(R)(\varepsilon^0) = \eta_a(R)(1)$ and $\eta(R)(\varepsilon^1) = \eta_m(R)(1)$. The argument involving a series of small checks shows that $\mathbf{Z}$ is an initial object in (rg_1), and so also in (crg_1), yields also the proof of the next proposition.

Proposition 1. The ring $k(\varepsilon)$ is an initial object in $rg_1(cc)$ and in $crg_1(cc)$.

We define connected rings by analogy with connected groups G in (cc), i.e., Hopf algebras, with the property that $k \to G$ induces an isomorphism $k \to G_0$ in degree zero for the initial object k. Note that the connected groups determine full subcategories $gr_0(cc)$ of gr(cc) and $ab_0(cc)$ of ab(cc).

Definition 3. A connected ring R in (cc) is a ring with unit such that $\eta(R)\colon k(\varepsilon) \to R$ induces an isomorphism $k(\varepsilon) \to R_0$ in degree zero.

We use the notations $rg_0(cc)$ and $crg_0(cc)$ for the full subcategories of $rg_1(cc)$ and $crg_1(cc)$, respectively, determined by connected rings. If $f\colon R \to R'$ is a morphism of connected rings, then $f\eta(R) = \eta(R')$.

Examples. (a) The direct sum and tensor product of vector bundles induce two H-space structures on BU. With these two structures $H_*(\mathbf{Z} \times BU, k)$ is a connected ring over any ground ring k. This example is considered further in Section 4, Example 1, and in Section 2.

(b) For a commutative ring A the maps

$$BGL_m(A) \times BGL_n(A) \to BGL_{m+n}(A)$$

induced by the direct sum of matrices and

$$BGL_m(A) \times BGL_n(A) \to BGL_{mn}(A)$$

induced by the tensor product collect to define two H-space structures on $\coprod_{1 \le n} BGL_n(A)$ such that $H_*(\Omega B(\coprod_{1 \le n} BGL_n(A)), k)$ is a connected ring. In Section 6 we will consider this example further for the case where A is a finite field.

2. Additive and Multiplicative Group Functors

The additive group functor $G_a\colon rg(\mathscr{X}) \to ab(\mathscr{X})$ is usually defined by the relation $G_a(R, \varphi_a(R), \varphi_m(R)) = (R, \varphi_a(R))$. Under certain circumstances we might want to modify this definition of G_a, which is the case, as we shall see, for connected rings.

The multiplicative group functor $G_m\colon rg_1(\mathscr{X}) \to gr(\mathscr{X})$ is more difficult to define in general, but basically it is the largest subgroup of the monoid $(R, \varphi_m(R), \eta_m(R))$, perhaps satisfying some additional condition. This article centers around the existence of an adjoint pair $R_g \dashv G_m\colon (rg_1(\mathscr{X}), gr(\mathscr{X}))$ for the category $\mathscr{X} = (cc)$, and so G_m is the basic functor for us. In this chapter we will only consider the existence of G_m and R_g for connected rings over (cc). The theory over (sets) is classical. Then we will proceed with topological applications.

For $\mathrm{rg}_0(\mathrm{cc})$ we wish to define G_a and G_m to have values in the categories of connected groups where the theory of Hopf algebras is more elementary. This will also suffice for most examples. To do this, we use the ring morphism $\varepsilon(R)\colon R \to k(\varepsilon)$ which is zero on R_i for $i > 0$ and satisfies $\varepsilon(R)\eta(R) = k(\varepsilon)$ where R is a connected ring in (cc). For a morphism $f\colon R \to R'$ of connected rings $\varepsilon(R')f = \varepsilon(R)$.

Definition 4. Let R be a connected ring in (cc).

(1) Using the morphisms $\varepsilon(R)$ and $k \to k(\varepsilon)$, where $1 \mapsto \varepsilon^0$ of coalgebras, we form the additive group

$$G_a(R) = k \,\square_{k(\varepsilon)}\, R$$

with the induced coalgebra structure and group structure coming from the restriction of $\varphi_a(R)$ to $G_a(R) \otimes G_a(R) \to G_a(R)$.

(2) Using the morphisms $\varepsilon(R)$ and $k \to k(\varepsilon)$ where $1 \mapsto \varepsilon^1$ of coalgebras, we form the multiplicative group

$$G_m(R) = k \,\square_{k(\varepsilon)}\, R$$

with the induced coalgebra structure and group structure coming from the restriction of $\varphi_m(R)$ to $G_m(R) \otimes G_m(R) \to G_m(R)$.

For $x \in R_0$ we see that $x \in G_a(R)$ if and only if $x = a\varepsilon^0$ some $a \in k$, and $x \in G_m(R)$ if and only if $x = a\varepsilon^1$ some $a \in k$. For $x \in R_i$ with $i > 0$ we have

$$x \in G_a(R) \text{ if and only if } \Delta x = x \otimes \varepsilon^0 + \varepsilon^0 \otimes x + \Sigma_j x_j \otimes x_j'$$
$$x \in G_m(R) \text{ if and only if } \Delta x = x \otimes \varepsilon^1 + \varepsilon^1 \otimes x + \Sigma_j x_j \otimes x_j'$$

where $0 < \deg(x_j)$, $\deg(x_j') < i$. From these formulas it is easy to see that the induced structures of $G_a(R)$ and $G_m(R)$ are well defined and that $G_m(R)$ is commutative if R is commutative. In addition, from the definition, we see that these are functors defined as follows:

$$G_a\colon \mathrm{rg}_0(\mathrm{cc}) \to \mathrm{ab}_0(\mathrm{cc}), \qquad G_m\colon \mathrm{rg}_0(\mathrm{cc}) \to \mathrm{gr}_0(\mathrm{cc})$$
$$G_m\colon \mathrm{crg}_0(\mathrm{cc}) \to \mathrm{ab}_0(\mathrm{cc}).$$

Now we have the following remarks relative to Examples (a) and (b) at the end of the previous section.

Remarks. (a) For the connected ring $H_*(\mathbf{Z} \times BU)$ we have

$$G_a H_*(\mathbf{Z} \times BU) = H_*(BU).$$

Moreover, the natural map of the H-spaces $P^\infty(\mathbf{C}) \to 1 \times BU \subset \mathbf{Z} \times BU$

commutes with the tensor product H-space structure on $\mathbf{Z} \times BU$ and hence defines a morphism in $\mathrm{ab}_0(\mathrm{cc})$

$$H_*(P^\infty(\mathbf{C}), k) \to G_m H_*(\mathbf{Z} \times BU, k).$$

In Section 4 we will see that this map satisfies a universal property. Recall that $H_*(P^\infty(\mathbf{C}), k) \xrightarrow{\sim} QH_*(BU, k)$ is a key step in the homological proof of Bott periodicity.

(b) For the connected ring $H_*(\Omega B(\coprod_{1 \le n} BGL_n(A)), k)$ we have $G_a H_*(\Omega B(\coprod_{1 \le n} BGL_n(A)), k) = H_*(BGL(A), k)$. Moreover, the natural map of the H-spaces $BGL_1(A) \to \Omega B(\coprod_{1 \le n} BGL_n(A))$ commutes with the tensor product H-space structure on $\Omega B(\coprod_{1 \le n} BGL_n(A))$ and hence defines a morphism in $\mathrm{ab}_0(\mathrm{cc})$

$$H_*(BGL_1(A), k) \to G_m H_*\left(\Omega B\left(\coprod_{1 \le n} BGL_n(A)\right), k\right).$$

In Section 5 we will see that this map satisfies a universal property in certain cases.

Case (a) is of course related to ordinary K-theory and case (b) to algebraic K-theory. To see the relation between the two formulations of these examples observe that $\mathbf{Z} \times BU$ and $\Omega B(\coprod_{1 \le n} BU_n)$ are isomorphic as homotopy ring objects.

3. Existence of the Group Ring Functor

There are two preliminaries needed for the construction of the group ring object. First, we use the functor S described in [1, Section 1,2], where $S(M) = \coprod_{0 \le n} S_n(M)$ and $S_n(M)$ is the quotient of $M^{n\otimes}$ by the action of the symmetric group permuting the factors. With the injection $\beta(M)\colon M = S_1(M) \to S(M)$ and the multiplication induced by $M^{m\otimes} \otimes M^{n\otimes} \to M^{(m+n)\otimes}$, $S(M)$ is the universal free commutative algebra on the graded module M. For a coalgebra C we can form $S(C)$ (or in the supplemented case $S(I(C))$) as in [1] and define a coalgebra structure compatible with the algebra structure by applying the universal factorization property to

$$C \to C \otimes C \to S(C) \otimes S(C).$$

This functor $C \mapsto S(C)$ satisfies a universal property which we summarize in the following proposition.

Proposition 2. The functor S: (cc) $\to$ ab(cc) is the left adjoint (or coadjoint) of the functor ab(cc) $\to$ (cc) which assigns to an abelian group its underlying coalgebra.

Hence S plays the role of the free abelian group functor, and it is what we will use to manufacture the group ring functor. Note, for a supplemented coalgebra C, as modules $C = k \oplus I(C)$, and hence as algebras $S(C) = S(k) \otimes S(I(C))$ where $S(k)$ is just the polynomial ring on one generator in degree 0.

Next, recall the switching morphism $T: L \otimes M \to M \otimes L$ is defined $T(x \otimes y) = (-1)^{pq} y \otimes x$ for $x \in L_p$, $y \in M_q$. This is just the switching morphism $T: C \otimes D \to D \otimes C$ defined because $C \otimes D$ is the product in (cc). By composing transpositions, we define a permutation morphism for each $\sigma \in \mathcal{S}_n$, the symmetric group

$$T_\sigma: M^{n\otimes} \to M^{n\otimes}.$$

The quotient $M^{n\otimes} \to S_n(M)$ considered above is just the largest quotient of $M^{n\otimes}$ on which all T_σ induce the identity.

For each factorization $n = ij$ of natural numbers we define $\sigma(i, j)$ to be the permutation of $\langle n \rangle = \{1, \ldots, n\}$ that is the following composite:

$$\langle n \rangle \xrightarrow{s(i,j)} \langle i \rangle \times \langle j \rangle \xrightarrow{t} \langle j \rangle \times \langle i \rangle \xrightarrow{s(i,j)^{-1}} \langle n \rangle,$$

where $t(a, b) = (b, a)$ and

$$s(i, j)(c) = \left(\left[\frac{c-1}{j} \right] + 1, c - \left[\frac{c-1}{j} \right] j \right).$$

As special cases

$$s(i, j)(1) = (1, 1), \ldots, s(i, j)(j) = (1, j), \ldots, s(i, j)(kj) = (k, j).$$

For $n = ij$ in the natural numbers we define

$$T_{i,j}: M^{n\otimes} \to M^{n\otimes}$$

by $T_{i,j} = T_{s(i,j)}$. Two special cases are $T_{1,1} = T$ and $T_{2,2} = X \otimes T \otimes X$. The latter was used in formulating the distributive law.

The *general distributive law* for a ring in (cc) is the assertion that the following diagram is commutative for all factorizations $n = ij$ by natural numbers where α is either $T_{j,i} \otimes 1$ or $1 \otimes T_{i,j}$:

$$(GDL) \qquad \begin{array}{ccccc} R^{i\otimes} \otimes R^{j\otimes} & \xrightarrow{\Delta_j(R)^{i\otimes} \otimes \Delta_i(R)^{j\otimes}} & R^{ij\otimes} \otimes R^{ij\otimes} & & \\ & & \downarrow{\scriptstyle \alpha} & & \\ \downarrow{\scriptstyle \varphi_a(R)_{(i)} \otimes \varphi_a(R)_{(j)}} & & R^{ij\otimes} \otimes R^{ij\otimes} & \xrightarrow{\varphi_m(R^{ij\otimes})} & R^{ij\otimes} \\ & & & & \downarrow{\scriptstyle \varphi_a(R)_{(ij)}} \\ R \otimes R & & \xrightarrow{\qquad \varphi_m(R) \qquad} & & R \end{array}$$

This is proved by double induction on i and j from the two cases $(i, j) = (2, 1)$ and $(1, 2)$ which correspond to just the ordinary left and right distributive laws.

Theorem 1. The functor G_m: $\mathrm{rg}_0(\mathrm{cc}) \to \mathrm{gr}_0(\mathrm{c}\ )$ has a left adjoint (or coadjoint) R_g: $\mathrm{gr}_0(\mathrm{cc}) \to \mathrm{rg}_0(\mathrm{cc})$. Moreover, the adjoint relation $R_g \dashv G_m$ restricts with the functor R_g to R_g: $ab_0(\mathrm{cc}) \to \mathrm{crg}_0(\mathrm{cc})$.

Proof. The second statement will follow easily from the first, and the construction of $R_g(G)$ will be made at the same time that we show there is a natural bijection

$$\mathrm{gr}(\mathrm{cc})(G, G_m(R)) \to \mathrm{rg}(\mathrm{cc})(R_g(G), R).$$

As an abelian group object $R_g(G) = k(\varepsilon) \otimes_{S(k)} S(G)$ in (cc), and as an algebra $R_g(G) = k(\varepsilon) \otimes S(I(G))$. As a morphism of coalgebras, $x \mapsto \varepsilon \otimes x$ is defined

$$G \gtrsim [k(\varepsilon) \otimes (k \oplus I(G))] \subset S(G) \subset [k(\varepsilon) \otimes SI(G)] = R_g(G)$$

and this composite will give rise to the adjunction morphism

$$\beta(G)\colon G \to G_m(R_g(G)).$$

For a morphism f: $G \to G_m(R)$ in $\mathrm{gr}_0(\mathrm{cc})$, we view it as a morphism in (cc) for the moment and apply Proposition 2 to find a morphism $\bar{f}$: $S(G) \to R$. The composite

$$\tilde{f} = \varphi_a(R)(\eta(R) \otimes_{S(k)} \bar{f})\colon R_g(G) = k(\varepsilon) \otimes_{S(k)} S(G) \to R$$

is well defined. To see this, restrict f to $f_{(1)}$: $I(G) \to R$ using $G_m(R) \subset R$ and define $f_{(i)}$: $S_i I(G) \to R$ by the requirement that the following diagram is commutative:

$$\begin{array}{ccc} I(G)^{i\otimes} & \xrightarrow{f^{i\otimes}} & R^{i\otimes} \\ \downarrow & & \downarrow \varphi_a(R)_{(i)} \\ S_i I(G) & \xrightarrow{f_{(i)}} & R \end{array}$$

Then $\tilde{f}$ satisfies the requirement that the following diagram is commutative for all i as graded modules:

$$\begin{array}{ccc} k(\varepsilon) \otimes S_i I(G) & \xrightarrow{\eta(R) \otimes f_{(i)}} & R \otimes R \\ \downarrow & & \downarrow \varphi_a(R) \\ R_g(G) = k(\varepsilon) \otimes SI(G) & \xrightarrow{\tilde{f}} & R \end{array}$$

The morphism $\tilde{\beta}(x) = \varepsilon \otimes x$ of coalgebras

$$G \rightleftarrows [k(\varepsilon) \otimes (k \oplus I(G))] \subset S(G) \subset [k(\varepsilon) \otimes_{S(k)} S(G)] = R_g(G)$$

is defined $\tilde{\beta}: G \rightarrow R_g(G)$. Since $\tilde{f}(\varepsilon) = \eta_m(R)(1)$, we see that $\tilde{f}\tilde{\beta}$ restricted from $G \rightarrow R$ to $G \rightarrow G_m(R)$ is just the given f.

Now referring to the generalized distributive law (GDL) above, we are led to define $\varphi_m(R_g(G))$ restricted to $S_i(G) \otimes S_j(G) \rightarrow S_{ij}(G)$ by the requirement that the following diagram is commutative:

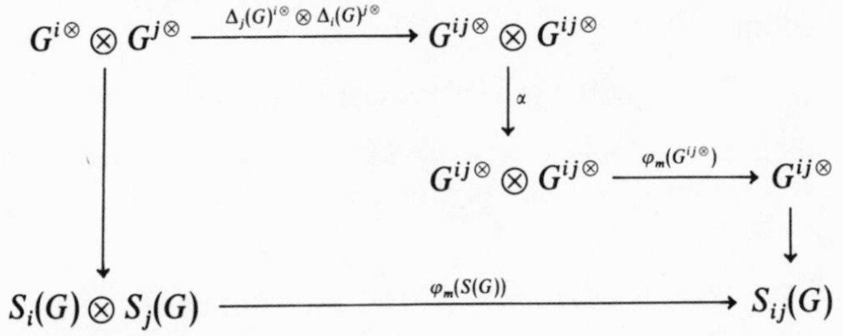

Now relating this to $f: G \rightarrow G_m(R)$ in $\mathrm{gr}^0(\mathrm{cc})$, we compare this diagram with (GDL) using the extension of $f: G \rightarrow R$ and the morphisms $f_{(i)}: S_i(G) \rightarrow R$ by the commutativity of the following diagram:

$$
\begin{array}{ccc}
G^{i\otimes} & \xrightarrow{f^{i\otimes}} & R^{i\otimes} \\
\downarrow & & \downarrow{\scriptstyle \varphi_a(R)_{(i)}} \\
S_i(R) & \xrightarrow{f_{(i)}} & R
\end{array}
$$

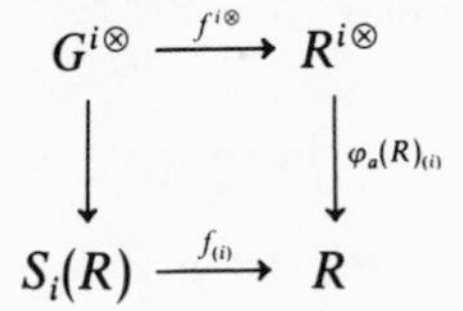

This comparison leads to the following commutative diagram:

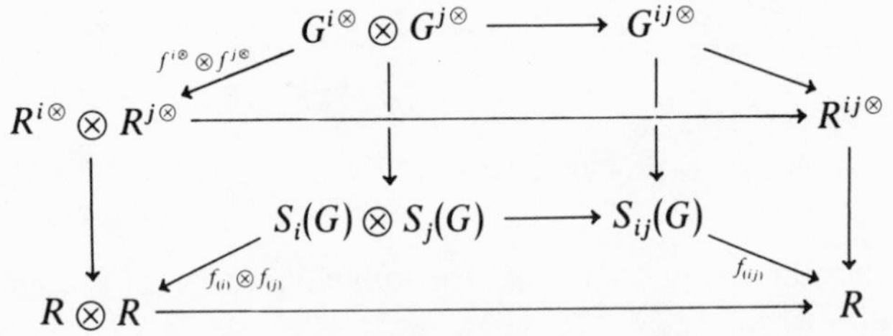

The top horizontal square is commutative because f is a morphism in $\mathrm{gr}_0(\mathrm{cc})$. Since the bottom horizontal square is a quotient of the top one, it is also commutative, and assembling this commutativity statement over all i and j, we deduce that $\tilde{f}: R_g(G) \rightarrow R$ is a morphism preserving the ring structure because we have only to check that in degree 0 where $S(G)_0 = S(k)$

maps correctly under $\tilde{f}$. But this is just the restriction of $\eta(R)\colon k(\varepsilon) \to R$ to $S(k) \subset k(\varepsilon)$.

Since the statements about commutative objects follow immediately, we have proven the theorem.

4. Basic Classes of Examples of Group Rings

In the previous section we used the functor S and now we consider also its dual S' described in [1, Section 1,2] where $S'(M) = \coprod_{0 \leq n} S_n'(M)$ and $S_n'(M)$ is the submodule of elements in $M^{n\otimes}$ fixed under the action of the symmetric group permuting the factors. With the projection $\alpha(M)$: $S'(M) \to S_1'(M) = M$ and the comultiplication induced by $M^{(m+n)\otimes} \to M^{m\otimes} \otimes M^{n\otimes}$, the coalgebra $S'(M)$ is the free commutative coalgebra on the graded module M. Note that $M_0 = 0$ if and only if $S'(M)$ is connected, which is the case we are primarily interested in.

For an algebra A we can form $S'(A)$ [or in the supplemented case $S'(I(A))$] as in [1] and define an algebra structure on $S'(A)$ [or $S'(I(A))$] compatible with the coalgebra structure by applying the universal factorization property to

$$S'(A) \otimes S'(A) \longrightarrow A \otimes A \xrightarrow{\varphi(A)} A$$

or

$$S'(I(A)) \otimes S'(I(A)) \longrightarrow I(A) \otimes I(A) \xrightarrow{I(\varphi(A))} I(A).$$

If $F\colon \mathrm{ab}_0(\mathrm{cc}) \to (\mathrm{ca})_0$ is the functor which assigns to a bicommutative connected Hopf algebra its underlying connected algebra, then the universal property of $S'I$ is contained in the next proposition (see [1]).

Proposition 3. The functor $S'I\colon (\mathrm{ca})_0 \to \mathrm{ab}_0(\mathrm{cc})$ is the right adjoint (or just adjoint) of $F\colon \mathrm{ab}_0(\mathrm{cc}) \to (\mathrm{ca})_0$.

We apply the functor $S'I$ to the case $A = k \oplus M$ where the product is zero on M, i.e., $M \cdot M = 0$. Thus $S'(I(k \oplus M)) = S'(M)$ has a bicommutative Hopf algebra structure. The underlying algebra $FS'(M)$ is sometimes denoted by $\Gamma(M)$ and it is the universal algebra with divided powers when M is a free module. The case of M of rank 1 is worked out in [1] and the general case can be built up from this one (see the Cartan Séminaire, 1954–1955).

Remark. Viewing $S'(M)$ in $\mathrm{ab}_0(\mathrm{cc})$ as above, we form as in Theorem 1 the group ring

$$R(M) = R_g(S'(M)) = k(\varepsilon) \otimes S(IS'(M)) = k(\varepsilon) \otimes_{S(k)} SS'(M),$$

which has a universal property relative to the adjunction morphism

$$\beta(M): S'(M) \to G_m R(M) = G_m R_g(S'(M))$$

in $\mathrm{ab}_0(\mathrm{cc})$. The groups $G_a R(M) = S(IS'(M))$ were studied in [1] where they were denoted $SS'(M)$, especially in the case where M was of rank 1.

In [1, Section 4] we defined a natural morphism

$$\lambda: SI(S'(M)) \to S'I(S(M)).$$

We denote the module free of rank 1 on a generator x in degree d by $[x, d]$. For $d \geq 1$ we showed in [1, 4.1] that for either d even, k a $\mathbf{Z}[\frac{1}{2}]$-algebra, or k an $\mathbf{F}_2$-algebra, the morphism $\lambda: SI(S'[x, d]) \to S'I(S[x, d])$ is an isomorphism and this Hopf algebra was denoted $B[x, d]$. Now by the above remark $G_a R[x, d] = B[x, d]$ and $R[x, d] = k(\varepsilon) \otimes B[x, d]$ as algebras. Now we can describe further the examples given in [1, 4.6] with more precision.

Example 1. The morphism $H_*(P^\infty(\mathbf{C}), k) \to G_m H_*(\mathbf{Z} \times BU, k)$ considered at the end of Section 2 is just the adjunction morphism $\beta[x, 2]: S'[x, 2] \to G_m R[x, 2] = k(\varepsilon^1) \otimes SI(S'[x, 2])$ in the above remark. This corresponds to a morphism of connected rings

$$R_g H_*(P^\infty(\mathbf{C}), k) \to H_*(\mathbf{Z} \times BU, k)$$

which is an *isomorphism*. To see this, recall that $\overline{H}_*(P^\infty(\mathbf{C}), k) \to QH_*(BU, k)$ is an isomorphism and $S\overline{H}_*(P^\infty(\mathbf{C}), k) \to H_*(BU, k)$ is surjective. By comparing ranks of the two sides, we see that $S\overline{H}_*(P^\infty(\mathbf{C}), k) \to G_a H_*(\mathbf{Z} \times BU, k)$ is an isomorphism from which we deduce the result.

The statement that $R_g H_*(P^\infty(\mathbf{C}), k) \to H_*(\mathbf{Z} \times BU, k)$ is an isomorphism is a very precise form of the splitting principle for characteristic classes of complex vector bundles on the universal example. Not only is the homology of $H_*(\mathbf{Z} \times BU, k)$ generated by the classes from $H_*(P^\infty(\mathbf{C}), k)$, but also the relations under permutations of sums of classes are included along with the multiplicative structure corresponding to tensor products of vector bundles. In the real case we have an example similar to the one corresponding to BU in the complex case.

Before going into further examples, we codify the principle used to show that $H_*(\mathbf{Z} \times BU)$ is a group ring object.

Proposition 4. For a morphism $u: G \to G_m(R)$ in $\mathrm{gr}_0(\mathrm{cc})$, the adjoint morphism $f: R_g(G) \to R$ is an isomorphism if and only if $\bar{u}: I(G) \to Q(G_a(R))$ is an isomorphism and $G_a(R) \cong S(L)$ for some L as an algebra.

Proof. The direct implication follows from the construction. For the converse observe that we have only to show that $G_a(f) = \bar{f}: G_a R_g(G) \to G_a(R)$ is an isomorphism since $R = k(\varepsilon) \otimes G_a(R)$ as algebras. Since $\bar{u}$ is an isomorphism, $\bar{f}: G_a R_g(G) \to G_a(R)$ is surjective. Since $G_a(R) \cong S(L)$ with $IG_a(R_g(G)) \leftrightarrows QS(L) = L$, there is a morphism $\bar{g}: G_a(R) \to G_a R_g(G)$ of algebras with $\bar{f}\bar{g}$ equal to the identity. Thus $Q(\bar{g})$ is the inverse of the isomorphism $Q(\bar{f})$. Hence $\bar{g}$ is surjective and $\bar{f}$ is injective, so f is an isomorphism. This proves the proposition.

Example 2. Let k be an $\mathbf{F}_2$-algebra. The morphism $H_*(P^\infty(\mathbf{R}), k) \to G_m H_*(\mathbf{Z} \times \mathrm{BO}, k)$ defined by the natural inclusion $P^\infty(\mathbf{R}) \to 1 \times BO \subset \mathbf{Z} \times BO$ is the adjunction morphism $\beta[x, 1]: S'[x, 1] \to G_m R[x, 1] = k(\varepsilon^1) \otimes SI(S'[x, 1])$ in the above remark. By Proposition 4 the adjunction morphism

$$R_g H_*(P^\infty(\mathbf{R}), k) \to H_*(\mathbf{Z} \times BO, k)$$

is an isomorphism.

Example 3. Let k be a $\mathbf{Z}[\frac{1}{2}]$-algebra. The composite map $P^\infty(\mathbf{C}) \to BU \to BO$ factors by $P^\infty(\mathbf{C})/(\pm 1) \to BO$, where the action of ± 1 on $P^\infty(\mathbf{C})$ is made free. Then the induced morphism $H_*(P^\infty(\mathbf{C})/(\pm 1), k) \to G_m H_*(\mathbf{Z} \times BO, k)$ is the adjunction morphism $\beta[x, 4]: S'[x, 4] \to G_m R[x, 4] = k(\varepsilon^1) \otimes SI(S'[x, 4])$ in the above remark. By Proposition 4 the adjunction morphism

$$R_g H_*(P^\infty(\mathbf{C})/(\pm 1), k) \to H(\mathbf{Z} \times BO, k)$$

is an isomorphism.

5. The Group Ring $H_*(\mathbf{Z} \times F\psi^q, k)$

Recall that $F\psi^q$ is the fibre of $\psi^q - 1: BU \to BU$ and $\mathbf{Z} \times F\psi^q$ is the homotopy-fixed point space of the action of $\psi^q: \mathbf{Z} \times BU \to \mathbf{Z} \times BU$. By a slight extension of the methods in [7, Section 1] $\mathbf{Z} \times F\psi^q$ admits a homotopy ring structure such that $\mathbf{Z} \times F\psi^q \to \mathbf{Z} \times BU$ perserves these ring operations up to homotopy. As the first step in the analysis of the ring $H_*(\mathbf{Z} \times F\psi^q, k)$, we recall the homology of $\mathrm{F}\psi^q$, and since it follows by the same method, also the homology of $J\psi^q$. For this, consider the following six-term fibre mapping sequence

$$U \longrightarrow F\psi^q \longrightarrow BU \xrightarrow{\psi^q - 1} BU \longrightarrow J\psi^q \longrightarrow U.$$

Over a field k of characteristic ℓ we use the induced fibre space and associated fibre bundle spectral sequences of Eilenberg and Moore, which collapse as in [2]. Hence

$$E^0H(F\psi^q) = \operatorname{Cotor}^{H_*(BU)}(k, H_*(BU))$$

and

$$E^0H(J\psi^q) = \operatorname{Tor}^{H_*(BU)}(k, H_*(BU)).$$

For $\ell = 0$ or for $\ell | q$, the homology $H_*(F\psi^q) = H_*(J\psi^q) = k$. For $\ell \neq 0$ or for $\ell \nmid q$, we denote by r the order of q in $\mathbf{F}_\ell^* \subset k^*$ and then,

$$E^0H_*(F\psi^q) = B[x, 2r] \otimes \bigotimes_{1 \leq m} E[y_m, 2rm - 1]$$

$$E^0H_*(J\psi^q) = B[x, 2r] \otimes \bigotimes_{1 \leq m} E[y_m, 2rm + 1].$$

These structures on E^0H_*, as a Hopf algebra, can be lifted to H_* as in [2], and in the case of $H_*(J\psi^q)$, each exterior generator y_n is a Bockstein of a generator in degree $2rn$. As objects in $\mathrm{ab}_0(\mathrm{cc})$, we have

$$H_*(\mathbf{Z} \times F\psi^q) = k(\varepsilon) \otimes B[x, 2r] \otimes \bigotimes_{1 \leq m} E[y_m, 2rm - 1]$$

and

$$H_*(J\psi^q) = B[x, 2r] \otimes \bigotimes_{1 \leq m} E[y_m, 2rm + 1].$$

In the next theorem we take up the question of $H_*(\mathbf{Z} \times F\psi^q)$ as a ring.

Theorem 2. For r the order of q in $\mathbf{F}_\ell^* \subset k^*$, there exists a morphism $u\colon S'[x, 2r] \otimes E[y, 2r - 1] \to G_m H_*(Z \times F\psi^q, k)$ in $\mathrm{ab}_0(\mathrm{cc})$ such that the adjoint morphism

$$R_g(S'[x, 2r] \otimes E[y, 2r - 1]) \to H_*(\mathbf{Z} \times F\psi^q, k)$$

is an isomorphism. Moreover, for $r = 1$, u is the homology morphism induced by a map

$$K(\mathbf{Z}/(q - 1), 1) \to 1 \times F\psi^q \subset \mathbf{Z} \times F\psi^q.$$

Proof. First we consider the case $r = 1$, which is equivalent to either $\ell | (q - 1)$ or $\bar{H}_*(\mathbf{Z}/(q - 1), 1; k) = I(S[x, 2] \otimes E[y, 1])$. The commutative diagram of spaces

$$\begin{array}{ccc}
F(\psi^q) = \ker(\psi^q) = K(\mathbf{Z}/(q-1), 1) & \longrightarrow & P^\infty \\
\downarrow & & \downarrow \\
\mathbf{Z} \times F\psi^q & \longrightarrow & \mathbf{Z} \times BU
\end{array}$$

induces the commutative diagram of graded modules with Bockstein operations

$$\begin{array}{ccc} \bar{H}_*(K(\mathbf{Z}/(q-1), 1) = I(S'[x, 2] \otimes E[y, 1]) & \to & I(S'[x, 2]) = \bar{H}_*(P^\infty) \\ \downarrow & & \downarrow \wr \\ QG_a H_*(\mathbf{Z} \times F\psi^q) & \longrightarrow & QG_a H_*(\mathbf{Z} \times BU). \end{array}$$

The horizontal morphisms are surjective, and the right-hand vertical morphism is an isomorphism. Hence $\gamma_k(x)$ maps injectively into $QG_a H_*(\mathbf{Z} \times F\psi^q)$. Since $\gamma_k(x)y$ is the Bockstein of $\gamma_{k+1}(x)$, it also maps nonzero to $QG_a H_*(\mathbf{Z} \times F\psi^q)$. So we deduce that $\bar{H}_*(K(\mathbf{Z}/(q-1), 1)) \to QG_a H_*(\mathbf{Z} \times F\psi^q)$ is an isomorphism, and the criterion in Proposition 4 applies to prove the theorem in the case $r = 1$.

For general $r \geq 1$ we cannot use $K(\mathbf{Z}/(q-1), 1)$ directly because $H_*(K(\mathbf{Z}/(q-1), 1)) = k$ for $r > 1$. Multiplication φ^q by q acts on $K(\mathbf{Z}/(q^r-1), 1)$ with order r and ψ^q acts on $F\psi^{q^r}$ with order r. These actions are made free and preserved by $K(\mathbf{Z}/(q-1), 1) \to F\psi^{q^r}$. We consider the following commutative diagram where the left-hand square is homotopy Cartesian and defines $F(\varphi^q, r)$:

$$\begin{array}{ccccc} F(\varphi^q, r) & \to & K(\mathbf{Z}/(q^r-1), 1) = F(\varphi^{q^r}) & \to & F(\varphi^{q^r})/(\varphi^q) \\ \downarrow & & \downarrow & & \downarrow \\ F\psi^q & \longrightarrow & F\psi^{q^r} & \longrightarrow & F\psi^{q^r}/(\psi^q) \end{array}$$

The action induced by φ^q or ψ^q in dimensions $2i-1$ or $2i$ on the homology is multiplication by q^i. Applying the covering space spectral sequence to a cyclic action of order r prime to ℓ (*note*: r divides $\ell - 1$), we observe that $H_*(F(\varphi^{q^r})/(\varphi^q)) = S'[\bar{x}, 2r] \otimes E[\bar{y}, 2r-1]$, the composite

$$H_*(F\psi^q) \to H_*(F\psi^{q^r}) \to H_*(F\psi^{q^r}/(\psi^q))$$

is an isomorphism since $H_*(F\psi^q) \to H_*(F\psi^{q^r})$ is a monomorphism, and $\bar{H}_*(F(\varphi^{q^r})/(\varphi^q)) \rightleftarrows QH(F\psi^{q^r}/(\psi^q))$. This implies that the comodule $H_*(F\psi^q)$ is an $H_*(F\psi^{q^r})$-injective, $H_*(F(\varphi^{q^r}))$ an $H_*(F(\varphi^{q^r})/(\varphi^q))$-injective, and $H_*(F\psi^{q^r})$ an $H_*(F\psi^{q^r}/(\psi^q))$-injective, being direct summands or extended comodules. Now we apply the induced fibre space spectral sequence and the change of coalgebra spectral sequence to deduce the first two isomorphisms, respectively:

$$\begin{aligned} H_*(F(\varphi^q, r)) &\rightleftarrows H_*(F\psi^q) \,\square_{H_*(F\psi^{q^r})}\, H_*(\varphi^{q^r}) \\ &\rightleftarrows H_*(F\psi^q) \,\square_{H_*(F\psi^{q^r}\,(\psi^q))}\, H_*(F(\varphi^{q^r})/(\varphi^q)) \\ &\rightleftarrows H_*(F(\varphi^{q^r})/(\varphi^q)). \end{aligned}$$

Hence the H-space $F(\varphi^q, r)$ mapping into $1 \times F\psi^q \subset \mathbf{Z} \times F\psi^q$ defines a map where $\overline{H}_*(F(\varphi^q, r)) \rightleftarrows QG_a H_*(\mathbf{Z} \times F\psi^q)$ is an isomorphism. The criterion of Proposition 4 applies and the adjoint morphism $R_g H_*(F(\varphi^q, r)) \to H_*(\mathbf{Z} \times F\psi^q)$ to the inclusion is the desired isomorphism. This proves the theorem.

Remark. In terms of the space $F(\varphi^q, r)$ constructed relative to the prime ℓ, in that r is the order of q in $F_\ell^* \subset k^*$, we can assert, in view of Theorem 2, that

$$F(\varphi^q, r) \to 1 \times F\psi^q \subset \mathbf{Z} \times F\psi^q$$

induces the group ring structure on $H_*(\mathbf{Z} \times F\psi^q)$. This situation is somewhat unsatisfactory in that $F(\varphi^q, r)$ depends on both q and r while $F\psi^q$ depends only on q.

6. The Group Rings $H_*(\mathbf{Z} \times BGL(\mathbf{F}_q), k)$ and $H_*(\mathbf{Z} \times BGL(\overline{\mathbf{F}}_q), k)$

Let q be a prime power p^a throughout this section where in the previous section it could be any natural number. Let $F = \mathbf{F}_q$ and $\overline{F}$ be the algebraic closure of F.

In [6] and [7] the Brauer lifting of representations in characteristic $p > 0$ to virtual representations in characteristic zero was used by Quillen to construct the following homotopy commutative diagram of H-spaces:

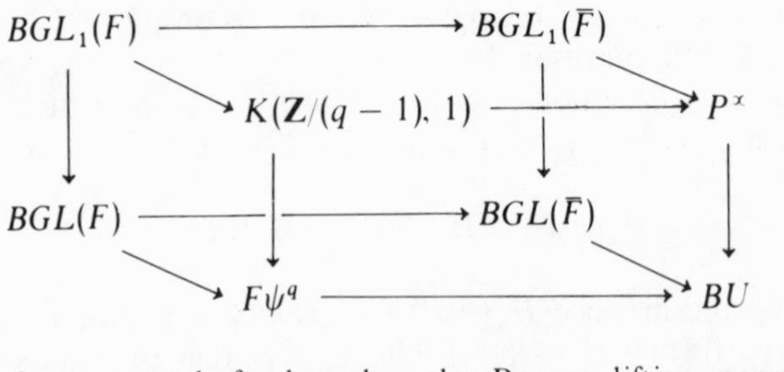

We have only to remark further that the Brauer lifting perserves both direct sum and tensor product structures and so is a morphism of homotopy rings.

Now in [6] Quillen proves that in characteristic $\ell \neq 0$ the following is an isomorphism

$$H_*(BGL(\bar{F})) \to H_*(BU),$$

and in [7] that in characteristic $\ell \neq p$ the following is an isomorphism

$$H_*(BGL(F)) \to H_*(F\psi^q).$$

Since we observed these are isomorphisms of rings, we are led to the following theorem:

Theorem 3. (1) The map $BGL_1(\bar{F}) \to 1 \times BGL(\bar{F}) \subset \mathbf{Z} \times BGL(\bar{F})$ induces a morphism $H_*(BGL_1(\bar{F})) \to G_m H_*(\mathbf{Z} \times BGL(\bar{F}))$ in $\mathrm{ab}_0(\mathrm{cc})$ such that the adjoint morphism

$$R_g H_*(BGL_1(\bar{F})) \to H_*(\mathbf{Z} \times BGL(\bar{F}))$$

is an isomorphism.

(2) For $r = 1$ or $\ell \mid (q-1)$, the map $BGL_1(F) \to 1 \times BGL(F) \subset \mathbf{Z} \times BGL(F)$ induces a morphism $H_*(BGL_1(F)) \to G_m H_*(\mathbf{Z} \times BGL(F))$ in $\mathrm{ab}_0(\mathrm{cc})$ such that the adjoint morphism

$$R_g H_*(BGL_1(F)) \to H_*(\mathbf{Z} \times BGL(F))$$

is an isomorphism.

(3) The ring $H_*(\mathbf{Z} \times BGL(F))$ admits the structure of a group ring.

Observe that by examples in Section 4 and Theorem 2, respectively, the following composites are isomorphisms independent of the calculations in Quillen's papers:

$$R_g H_*(BGL_1(\bar{F})) \to H_*(\mathbf{Z} \times BGL(\bar{F})) \to H_*(\mathbf{Z} \times BU),$$

and for $r = 1$,

$$R_g H_*(BGL_1(F)) \to H_*(\mathbf{Z} \times BGL(F)) \to H_*(\mathbf{Z} \times F\psi^q).$$

To show that $R_g H_*(BGL_1(\bar{F})) \to H_*(\mathbf{Z} \times BGL(\bar{F}))$ and $R_g H_*(BGL_1(F)) \to H_*(\mathbf{Z} \times BGL(F))$ are also surjective, and hence isomorphisms, we have only to apply the following consideration also used by Quillen in his two papers cited above.

Definition. The mod ℓ homology of a finite group G is detected by abelian ℓ^a-subgroups ($a \geq 1$) provided

$$\coprod_A H_*(A) \to H_*(G)$$

is surjective where A runs over the class of abelian subgroups of G such that $\ell^a A = 0$.

Quillen's Detection Theorem. If the mod ℓ homology of G is detected by abelian ℓ^a-subgroups, then the mod ℓ homology of the semidirect product $\mathcal{S}_n \propto G^n$ is also detected by abelian ℓ^a-subgroups.

This applies to $BGL_n(F)$ because for $n = rm + e, 0 \leq e < r, \mathcal{S}_m \propto GL_r(F)^m \subset GL_n(F)$ contains an ℓ-Sylow subgroup of $GL_n(F)$ where r is the order of q in $\mathbf{F}_\ell^*$. For $q^r - 1 = \ell^a c$, where ℓ does not divide c, the ℓ-Sylow subgroup of $GL_r(\mathbf{F}_q)$ is contained in $\mathbf{F}_{q^r}^*$ and is cyclic of order ℓ^a, where $\mathbf{F}_{q^r} \subset GL_r(\mathbf{F}_q)$, is the totally nonsplit torus.

Similar, but more complicated, considerations hold for $r > 1$, and this leads to a different perspective on the very important results in [6] and [7].

REFERENCES

[1] D. Husemoller, The structure of the Hopf algebra $H_*(BU)$ over a $\mathbf{Z}_{(p)}$-algebra, *Amer. J. Math.* **93** (1971), 329–349.

[2] D. Husemoller, On the homology of the fibre of $\psi^q - 1$, *Proc. Seattle Algebraic K-theory Conf. I.* Springer Lecture Notes, No. 341 (1973).

[3] S. Mac Lane, "Categories for the Working Mathematician," Springer-Verlag, Berlin and New York, 1970.

[4] R. J. Milgram, The mod 2 spherical characteristic classes, *Ann. of Math.*, **92**, (1970), 238–261.

[5] J. Milnor and J. C. Moore, On the structure of Hopf algebras, *Ann. of Math.* **81** (1965), 211–264.

[6] D. Quillen, The Adams conjecture, *Topology* **10** (1971), 67–80.

[7] D. Quillen, On the cohomology and K-theory of the general linear groups over a finite field, *Ann. of Math.* **96** (1972), 552–586.

While this paper was being written, the author was a guest of the I.HE.S. in Bures sur Yvette and the S.F.B. in Bonn, and he would like to acknowledge their hospitality. During the year 1974–1975 we were also receiving sabatical support from Haverford College.

AMS 55E10, 55E50, 55E15

DEPARTMENT OF MATHEMATICS
HAVERFORD COLLEGE
HAVERFORD, PENNSYLVANIA

A Whitehead Theorem

D. M. KAN

1. Introduction

The classical Whitehead theorem states:

If $f: X \to Y$ is a map between simply connected spaces such that $H_* f$ is an isomorphism for $i \leq n$ and an epimorphism for $i = n + 1$, then $\pi_i f$ is also an isomorphism for $i \leq n$ and an epimorphism for $i = n + 1$.

In this note we will use some ideas of A. K. Bousfield and E. Dror to show that one can remove the assumption that X and Y are simply connected if one changes the conclusion as follows:

For a map $f: X \to Y$ between connected spaces one can consider all factorizations

$$X \to X' \to Y$$

in which the map $X \to X'$ is a cofibration and induces an isomorphism $H_* X \approx H_* X'$. Bousfield [1] has shown that there are terminal ones among these factorizations and that any two terminal ones are homotopically equivalent. He also constructed a functorial such terminal factorization

$$X \to Ef \xrightarrow{f'} Y$$

Our Whitehead theorem now becomes:

If $f: X \to Y$ is a map between (pointed) connected spaces such that $H_i f$ is an isomorphism for $i \leq n$ and an epimorphism for $i = n + 1$, then $\pi_i f'$ is also an isomorphism for $i \leq n$ and an epimorphism for $i = n + 1$.

Moreover this result also holds for homology with coefficients in *the integers modulo a prime p* or *a subring of the rationals*.

Notation and Terminology. Throughout this note H_* will denote *homology with coefficients in the integers modulo a prime p* or *a subring of the rationals.*

We will work in the category $\mathscr{S}_*$ of *pointed simplicial sets* and refer the reader to [3] and [2, Ch. VIII] for more details on simplicial sets and on their relation to topological spaces.

2. The Bousfield Factorization

In this section we briefly review the main result of [1]. First we recall the definition of

2.1. H_*-Fibrations. A map $u: X \to Y \in \mathscr{S}_*$ is called an *H_*-fibration* if it has the right lifting property with respect to every map $i: A \to B \in \mathscr{S}_*$ that is an injection and induces an isomorphism $H_* A \approx H_* B$, i.e., for every commutative square

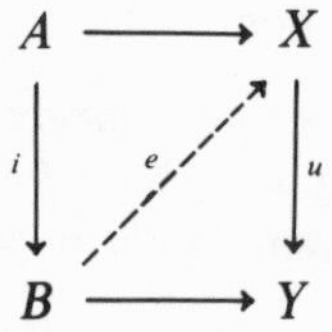

there exists a map e that makes the triangles commute.

Clearly *every H_*-fibration is a fibration.*

Bousfield's [1] main result then is the

2.2. Bousfield Factorization Theorem [1, 11.1]. *For every map $f: X \to Y \in \mathscr{S}_*$ there is a natural factorization*

$$X \to Ef \xrightarrow{f'} Y \in \mathscr{S}_*$$

in which

(*i*) *the map $X \to Ef$ is an injection and induces an isomorphism $H_* X \approx H_* Ef$, and*

(*ii*) *the map $f': Ef \to Y$ is an H_*-fibration.*

Moreover it is not hard to see that the properties 2.2(i) and 2.2(ii) completely determine the homotopy type of Ef, i.e., one has

2.3. Proposition. *Let $f: X \to Y \in \mathscr{S}_*$ and let*

$$X \to X' \to Y \in \mathscr{S}_*$$

be a factorization of f in which

(*i*) *the map $X \to X'$ satisfies 2.2(i), and*
(*ii*) *the map $X' \to Y$ satisfies 2.2(ii).*

Then there is a commutative diagram

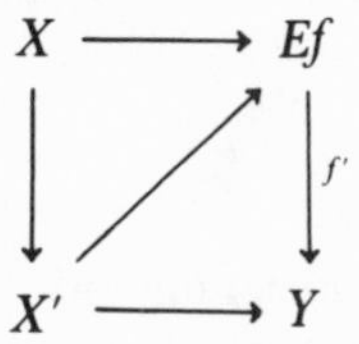

in which the map $X' \to Ef$ is a weak homotopy equivalence.

3. The Whitehead Theorem

We now state our

3.1. Whitehead Theorem. *Let $n \geq 0$ and let $f: X \to Y \in \mathscr{S}_*$ be such that X and Y are connected and that $H_i f$ is an isomorphism for $i \leq n$ and an epimorphism for $i = n + 1$. Then, in the notation of 2.2, $\pi_i f'$ is also an isomorphism for $i \leq n$ and an epimorphism for $i = n + 1$.*

The key to the proof is the following

3.2. Lemma. *Let $f: X \to Y \in \mathscr{S}_*$ be as in 3.1. Then f admits a factorization*

$$X \to X' \xrightarrow{g} Y$$

in which

(i) *the map $X \to X'$ is an injection and induces an isomorphism $H_* X \approx H_* X'$, and*
(ii) *$\pi_i g$ is an isomorphism for $i \leq n$ and an epimorphism for $i = n + 1$.*

Proof of 3.1. Applying 3.2 to the map f': $Ef \to Y$ one gets a factorization

$$Ef \to E' \xrightarrow{g} Y$$

and the desired result now follows immediately from the existence of the commutative diagram

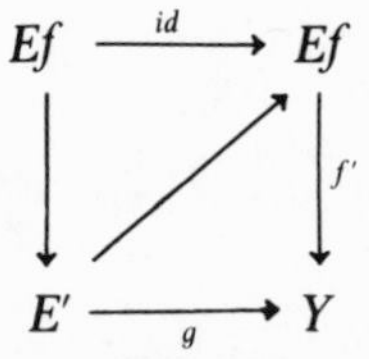

We end with a

Proof of 3.2. It is clearly possible to construct a factorization of f

$$X \to X''' \to X'' \to Y' \to Y \in \mathscr{S}_*$$

in which

(i) the map $X \to Y'$ is an injection and the map $Y' \to Y$ is a fibration and a weak homotopy equivalence,

(ii) the map $X''' \to Y'$ maps X''' isomorphically onto the union of the $(n+1)$-skeleton of Y' and the image of X under the map $X \to Y'$, and

(iii) the map $X''' \to X''$ is an injection and a weak homotopy equivalence and the map $X'' \to Y'$ is a fibration.

Then the induced map $\pi_i X'' \to \pi_i Y$ is an isomorphism for $i \leq n$ and is an epimorphism for $i = n+1$. Furthermore $H_i(X'', X) = 0$ for $i \neq n+1$ and $H_{n+1}(X'', X)$ is a free R-module, if R is the coefficient ring. If F denotes the fibre of the fibration $X'' \to Y$, then one has a commutative diagram

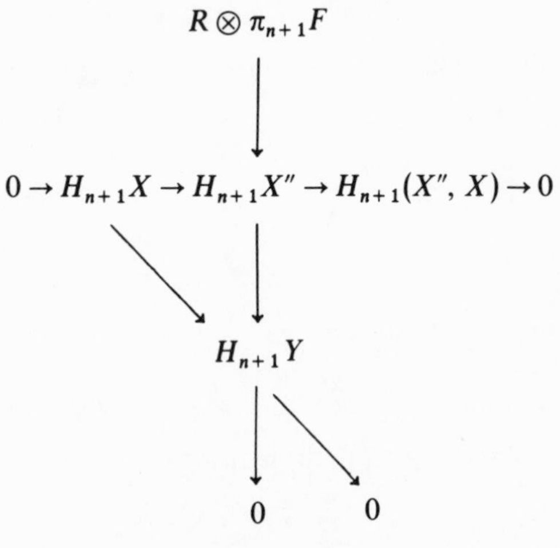

that is exact in all three directions. This implies that the map $R \otimes \pi_{n+1}F \to H_{n+1}(X'', X)$ is onto. Thus one can choose a set of spherical $(n+1)$-simplices in F that go to a free R-basis of $H_{n+1}(X'', X)$ and X' is now obtained from X'' by "attaching an $(n+2)$-simplex to each of these spherical $(n+1)$-simplices."

REFERENCES

[1] A. K. Bousfield, The localization of spaces with respect to homology, *Topology* **14** (1975), 133–150.

[2] A. K. Bousfield and D. M. Kan, Homotopy limits, completions and localizations, "Lecture Notes in Mathematics," Vol. 304. Springer-Verlag, Berlin and New York, 1972.

[3] J. P. May, "Simplicial Objects in Algebraic Topology." Van Nostrand-Reinhold, Princeton, New Jersey, 1967.

The author was partially supported by the National Science Foundation.

AMS 55D10

DEPARTMENT OF MATHEMATICS
MASSACHUSETTS INSTITUTE OF TECHNOLOGY
CAMBRIDGE, MASSACHUSETTS

Variable Quantities and Variable Structures in Topoi

F. WILLIAM LAWVERE

In memory of my eldest son, William Nevin.

I have organized this chapter into three sections as follows:

1. The conceptual basis for topoi in mathematical experience with variable sets.
2. A formal theory of variable abstract sets as a relativized foundation for geometry and analysis, with due attention to "the" case of constant sets.
3. Sheaves of continuous maps, étendues, and a proposed distinction between variable quantities in particular and variable structures in general.

Readers of Section 1 who are not too familiar with recent work on topoi may find clarification of some concepts in Section 2. Section 3 treats two aspects of sheaf theory not yet sufficiently incorporated into general topoi theory, with some remarks on the possible relevance of their relation to analysis and philosophy.

1.

Around 1963 (the same year in which I completed my doctoral dissertation under Professor Eilenberg's direction) five distinct developments in geometry and logic became known, the subsequent unification of which has, I believe, forced upon us the serious consideration of a new concept of set. These were the following:

"Non-Standard Analysis" (A. Robinson)
"Independence Proofs in Set Theory" (P. J. Cohen)
"Semantics for Intuitionistic Predicate Calculus" (S. Kripke)
"Elementary Axioms for the Category of Abstract Sets" (F. W. Lawvere)
"The General Theory of Topoi" (J. Giraud)

Apart from these specific developments, there has long been in geometry and differential equations the idea that the category of families of spaces smoothly parametrized by a given space X is similar in many respects to the category of spaces itself, and indeed, from the point of view of physics, it is perhaps to such a category with X "generic" or unspecified that our stably correct calculations refer, since there are always small variations or further parameters that we have not explicitly taken into account; the "new" concept of set is in reality just the logical extension of this idea. Of the five specific developments referred to, the decisive one for the concept of variable set was the theory of topoi; while nonstandard analysis, the forcing method in set theory, and Kripke semantics all involved, as will be explained below, sets varying along a poset $\mathbf{X}$, it was Grothendieck, Giraud, Verdier, Deligne, M. Artin, and Hakim who, by developing topos theory, made the qualitative leap—well-grounded in the developments in complex analysis, algebraic geometry, sheaf theory, and group cohomology during the 1950s—to consideration of sets varying along a small category $\mathbf{X}$ and at the same time emphasized that the fundamental object of study is the whole category of sets so varying. Those insisting on formal definitions may thus, in what follows, consider that "variable set" simply means an object in some (elementary) topos (just as, using an effective axiom system to terminologically invert history, we sometimes say that "vector" means an element of some vector space).

Traditionally, set theory has emphasized the constancy of sets, and both Robinson's nonstandard analysis and Cohen's forcing method involve passing from a system $\mathscr{S}$ of supposedly constant sets to a new system $\mathscr{S}'$ that still satisfies the basic axioms for constant sets; however, it is striking that both methods pass "incidentally" through systems of variable sets, and further, that the distinction between the two methods lies in the distinction between two fundamental ways of analyzing variation.

Let us recall what these two ways of analyzing variation are, first in the case of variable quantity. Both involve separating a *domain* of variation and a *type* of quantity (discrete, continuous, scalar, vector, tensor, operator, functional, etc.); let us fix the case R of continuous scalar quantity since the basic distinction between the two analyses is in how the domain of variation is treated. According to the first analysis, the domain X consists of "points" (points of space, instants of time, particles of a body, etc.), and a variable quantity is identified with a mapping $X \to R$; conditions such as continuity or measurability of the variation have to be imposed as additional properties involving additional structure on X. Although the foregoing is the usual view, it is not always adhered to in practice; for example, if X, μ is a measure space, then any member f of the usual $L_p(X, \mu)$ is clearly a variable quantity with domain of variation X, although it makes no sense to speak of the value of f at a point x. The second analysis is to consider that the domain $\mathbf{X}$ consists of parts (subregions of space, subintervals of time, parts of a body, etc.) and that a variable quantity is identified with a lattice homomorphism from parts of R into parts of $\mathbf{X}$; this remains sensible even if parts of $\mathbf{X}$ with μ-null difference are regarded as indistinguishable. The first analysis may be considered as a special case of the second by considering $\mathbf{X} = 2^X$. Conversely, if $\mathbf{X}$ is a complete Heyting algebra we can define its points as the infinitary sup-preserving lattice homomorphisms $\mathbf{X} \to 2$, which if $\mathbf{X} = 2^X$, or more generally if $\mathbf{X}$ is a sober topology for X, will correspond exactly to the mappings $1 \to X$, i.e., to points in the usual sense; of course if $\mathbf{X}$ is measurable sets modulo null sets there often will not be any points. We may also consider as *ideal points* the *finitary* lattice homomorphisms $\mathbf{X} \to 2$, which in case $\mathbf{X} = 2^X$ are just the ultrafilters on X; i.e., the ideal points are the points of the compactification, and the axiom of choice tries to reassure us that at least ideal points exist for any $\mathbf{X}$.

Returning now to nonstandard analysis and forcing, we start with a model $\mathscr{S}$ of a theory of constant sets. If X is a given (say countable) constant set, then $\mathscr{S}^X$ (i.e., all functions from X to $\mathscr{S}$) is a system of variable sets (conforming to the first analysis of variation). If we "stop" the variation at a point of X, we of course get back $\mathscr{S}$; but if we stop ("localize") the variation at an *ideal point* in $\beta(X) - X$, we get a new system $\mathscr{S}'$ of sets that satisfies the same elementary axioms (e.g., those expressing constancy) as $\mathscr{S}$, but which will definitely be different from $\mathscr{S}$. In particular, it will contain new "infinitesimal" elements—the residual traces of the variation that has been "stopped"—shown by Robinson to permit a reduction in the complexity of many definitions and proofs in analysis. For the forcing method we need, however, the second description of variation, applied to variable sets rather than variable quantities; instead of a set X we need a poset $\mathbf{P}$ in $\mathcal{S}$. As was later clarified by Scott and Solovay, it is more invariant though sometimes less convenient to enlarge $\mathbf{P}$ to the Boolean algebra $\mathbf{X}$ of $\neg\neg$-stable elements of the Heyting algebra of all order-preserving maps $\mathbf{P} \to 2$; this $\mathbf{X}$ typically has no points, but we can still consider sets E, which vary along it as follows: for each $A \in \mathbf{X}$, $E(A) \in \mathcal{S}$, and whenever $A \subseteq B$ in $\mathbf{X}$, there is a restriction mapping $E(B) \to E(A)$, and these fit together functorially whenever $A \subseteq B \subseteq C$; moreover, there is the condition

$$E(A) \simeq \prod_{i \in I} E(A_i)$$

whenever $A = \sum_{i \in I} A_i$ is a *disjoint* union in $\mathbf{X}$. If E_1, E_2 are as just described, a "mapping" $E_1 \xrightarrow{\varphi} E_2$ means any family $E_1(A) \xrightarrow[\varphi A]{} E_2(A)$ of mappings in $\mathcal{S}$ indexed by the A in $\mathbf{X}$ and satisfying the commutativity

$$\begin{array}{ccc} E_1(B) & \xrightarrow{\varphi B} & E_2(B) \\ \downarrow & & \downarrow \\ E_1(A) & \xrightarrow{\varphi A} & E_2(A) \end{array} \qquad \text{whenever } A \subseteq B \text{ in } \mathbf{X}$$

Thus we have defined a category $\mathcal{E}$ of sets varying along $\mathbf{X}$ (usually called Boolean-valued sets). Below we will see that it is not necessary to define the traditional ε-relation in $\mathcal{E}$: in any case, the interesting set-theoretic questions such as choice, replacement, the continuum hypothesis, measurable cardinals, etc. are categorical invariants anyway, that is, the questions depend only on how maps compose, not on an a priori notion of iterated membership. We can again localize at any chosen ideal point of $\mathbf{X}$ to obtain $\mathcal{S}'$, a new system of sets that look constant insofar as the most elementary properties [such as axiom of choice, two valuedness (see below)] that distinguish constant from variable sets are concerned, but unlike the previous case of nonstandard analysis, some of the deeper properties that had been proposed to enforce constancy, such as the axiom of constructibility (by taking as $\mathbf{P}$ the basic open sets of the Cantor space) or the continuum hypothesis (by taking $\mathbf{P}$ to be the basic open sets of a big generalized Cantor space), are as Cohen showed destroyed by the passage $\mathcal{S} \rightsquigarrow \mathcal{S}'$ even though "elementary" in the technical sense. It was these examples $\mathcal{S}'$ that led Tierney and me to further generalize the previous theory of topoi in 1969 by making it elementary, since although $\mathcal{S}$, $\mathcal{S}^X$, $\mathcal{E}$ are topoi in the Grothendieck–Giraud sense, the $\mathcal{S}'$ of nonstandard analysis and the $\mathcal{S}'$ of forcing are not; on the other hand, the essential issues of nonstandard analysis and of forcing can be dealt with in a perhaps more natural and certainly more invariant fashion in $\mathcal{S}^X$, respectively $\mathcal{E}$, provided one does not insist on constancy (which here just means on two-valuedness since the axiom of choice and hence the law of excluded middle are already valid in $\mathcal{S}^X$ and $\mathcal{E}$ if they are in $\mathcal{S}$).

Variable sets arose only "incidentally" in nonstandard analysis and forcing. The original goal was to construct new models of the theory of constant sets bearing nontrivial relation to a given such model. By contrast, in Kripke semantics for the Heyting predicate calculus the variation is essential also in the end result. Indeed, the thrust of Kripke's completeness theorem is that no logic stronger than intuitionistic logic can be valid for sets that are varying in any serious way, and in the other direction the Heyting predicate calculus *is* valid in *all topoi*, although topoi are qualitatively more general in at least two ways than the models for that calculus considered by Kripke in 1963. The latter also involved a system of variable sets $\mathcal{S}^{\mathbf{P}^{op}}$ identified with sets varying along a poset $\mathbf{P}$ according to the second analysis of variation, but more simply than with forcing and Boolean-valued sets as described above. We only consider $E(A) \in \mathcal{S}$ for $A \in \mathbf{P}$ itself and the transition mapping $E(B) \to E(A)$ in $\mathcal{S}$ whenever $A \leq B$ in $\mathbf{P}$; the transitions are subject to functoriality (transitivity), but to no further conditions, and the mappings $E_1 \to E_2$ in $\mathcal{S}^{\mathbf{P}^{op}}$ are defined as before. The interpretation of these variable sets was in terms of subjective variation of knowledge; the elements of $\mathbf{P}$ are called stages of knowledge and $A \leq B$ is taken to mean that A is a deeper (or later) stage of knowledge than B; for any set E we have, at any given stage B, constructed certain elements of E and proved certain equalities between pairs of elements constructed, giving an abstract set $E(B)$; if $A \leq B$ is a deeper stage of knowledge, the transition map $E(B) \to E(A)$ reflects that no constructed elements are ever lost and no proven equations are ever disproved, but the map is neither surjective nor injective since new elements may be constructed and new equalities proved at stage A. Considering that an n-ary relation S on E means simply another variable set equipped with a monomorphic mapping $S \rightarrowtail E^n$ in $\mathcal{S}^{\mathbf{P}^{op}}$, where E^n is the cartesian power in $\mathcal{S}^{\mathbf{P}^{op}}$, one finds that the operation of substitution is easily defined and that the operations of conjunction, disjunction, implication, and universal and existential quantification on relations, which are uniquely defined by the rules of inference, exist in $\mathcal{S}^{\mathbf{P}^{op}}$. The crucial fact is that universal quantification and implication do not commute with evaluation at a stage, but rather

$$\begin{gathered}\forall x\,[S_1(x, y) \Rightarrow S_2(x, y)] \text{ holds at } B, \text{ where } y \in E(B)\\ \text{iff for all } A \text{ with } A \leq B \text{ and for all } x \in E(A)\\ \text{if } S_1(x, y|A) \text{ then } S_2(x, y|A)\end{gathered}$$

so that such a relation's truth value at B depends on its classical truth value at all deeper stages. Since by definition $\neg S = [S \Rightarrow \text{false}]$, a similar situation holds for logical negation, so that in particular $S \Rightarrow \neg\neg S$ but not conversely. Though an infinite $\mathbf{P}$ is required to simultaneously refute all intuitionistically nonprovable inferences of predicate calculus, the two-element poset

$$\mathbf{P} = \{U \leq X\} = \mathbf{2}$$

suffices to refute the inference from $\neg\neg S$ to S; moreover there is an immediate connection with geometry since for $\mathbf{P} = \mathbf{2}$ it is easily verified that $\mathscr{S}^{\mathbf{P}^{\mathrm{op}}}$ is equivalent to the category of $\mathscr{S}$-valued sheaves on a two-point topological space with three open sets. As in any topos there is a unique "set" Ω of truth values in $\mathscr{S}^{\mathbf{P}^{\mathrm{op}}}$, which in the case $\mathbf{P} = \mathbf{2}$ is just the two-stage variable set

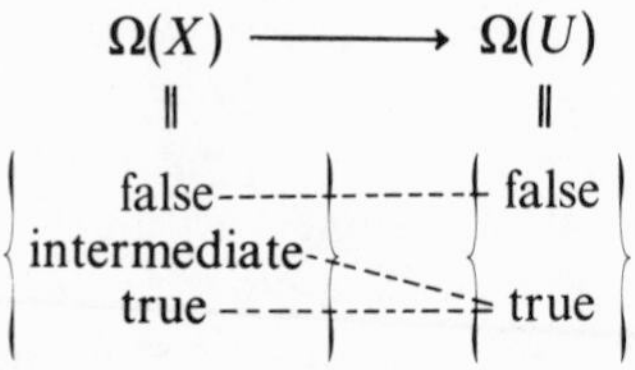

the intermediate value reflecting the fact that for an inclusion $S \rightarrowtail E$ in $\mathscr{S}^{\mathbf{2}^{\mathrm{op}}}$, there may be elements of $E(X)$ that are not in $S(X)$ but that do get mapped into $S(U)$ upon "restriction" to $E(U)$. Then $(\neg\neg S)(X)$ consists of all the elements of $E(X)$ that on restriction are in $S(U)$, so that $S \subseteq \neg\neg S$ is in general a proper inclusion. I would like to emphasize that recognizing the central importance for mathematics of the Heyting predicate calculus (i.e., intuitionistic logic) in no way depends on accepting a subjective idealist philosophy such as constructivism; objectively variable sets occur (at least implicitly) every day in geometry and physics and the fact that this variation is reflected in our minds in no way means that it is "freely created" by our minds; but it seems to have been the intuitionists who first succeeded in formulating the logic that holds for at least a certain definite portion of variation in general.

Some idea of which portion may be conveyed by the following example. Suppose the variation is along the temporal ordering and consider the statement:

The feudal landlords are the ruling class.

Recalling that truth is preserved by the transition maps in the sense that once something is true, it remains true, we see that the above statement is not an acceptable relation S in such a hypothetical temporal topos since it was once false, became true for a period, then became false again. On the other hand, consider the statement:

The feudal landlords have ruled.

Although this statement was sometimes true and sometimes false it has, solely by virtue of its grammatical structure, the requisite property that as soon as it became true, it remained forever true. Of course there are also more profound ways of dealing with temporal variation than simply identifying time with a poset that is governing the variation in the simple way we have suggested here, and some of these can even be accounted for by a suitable topos.

In Kripke's topoi existential quantification and disjunction *do* commute with evaluation at stages:

$$\exists x\, S(x, y) \text{ is true at } B \text{ for } y \in E(B) \quad \text{iff} \quad \text{there is } x \text{ in } E(B) \text{ with } S(x, y).$$

$$S_1(x) \vee S_2(x) \text{ is true at } B \quad \text{iff} \quad S_1(x) \text{ is true at } B \text{ or } S_2(x) \text{ is true at } B.$$

To put it more set-theoretically, images of mappings and unions of subsets commute with evaluation at B. But these facts apparently cannot be maintained in modeling Heyting-type theory (i.e., intuitionistic analysis) even though the Kripke topoi do have an intrinsic higher-order structure. At any rate, they certainly are false in the kind of topos that was already well understood in the 1950s, namely in a category of set-valued sheaves on a topological space. Such is also governed by a poset, namely the complete Heyting algebra of open sets of a space, but all the variable sets E satisfy the familiar "pasting" or sheaf condition, which is similar to but more involved than the infinite product condition for *disjoint* coverings mentioned above in connection with Boolean-valued sets. In particular, the image of a map between variable sets is also required to be a sheaf, which has the following effect. Suppose $E_1 \xrightarrow{f} E_2$ is a map of sheaves and $y \in E_2(U)$ for some open set U in the space over which the sets are varying. Then the rules of inference force

> $\exists x[f(x) = y]$ is true on U iff there exists an open covering U_i of U and there exist $x_i \in E_i(U_i)$ such that $f(x_i) = y | U_i$, $i \in I$,

where $y | V \in E(V)$ denotes the restriction of y to $V \subseteq U$. A similar statement holds for disjunction (union of two subsheaves) and these facts (with "covering" suitably interpreted) hold in any topos.

Both the Kripke topoi and the topological topoi are generated by their subsets of 1 in the sense that

> if $S \rightarrowtail E$ is any monomorphism, then either it is an isomorphism or else there exists an object U for which $U \to 1$ is a monomorphism and there exists a morphism $U \xrightarrow{x} E$ that does not factor through S (i.e., $x \notin S$).

Examples of topoi of variable sets for which this condition does not hold were known implicitly for a long time and more explicitly in the 1950s in connection with group cohomology, in the form of the category $\mathcal{S}^G$ of G-sets (permutation representations) for a group G. Here the only subsets U of 1 are 0 and 1, yet a map $1 \xrightarrow{x} E$ is only an element of E that *is fixed by* G; or looking at it from the other side, the most natural object that is the source of enough elements x (as U in the above condition) is G itself acting on itself by translation, yet (if $G \neq 1$) it is not a subset of 1. Now upon taking abelian-group-object categories, we have for a space X

$$\text{Ab}(\text{Sheaves}(X, \mathcal{S})) = \text{abelian sheaves on } X$$

and for a group G

$$\text{Ab}(\mathcal{S}^G) = G\text{-modules}.$$

Moreover, the functor represented by 1 [i.e., $\mathcal{X}(1, \_)$] becomes, for $\mathcal{X} = \text{Sheaves}(X, \mathcal{S})$, the global sections functor

$$\text{Sheaves}(X, \mathcal{S}) \xrightarrow{\Gamma} \mathcal{S}$$

and, for $\mathcal{X} = \mathcal{S}^G$, the fixed point functor

$$\mathcal{S}^G \to \mathcal{S},$$

which upon taking Ab and taking right-derived functors via injective resolutions as in Cartan–Eilenberg, become respectively

$$H^n(\mathcal{X}, E) = H^n(X, E), \qquad \text{and} \qquad H^n(\mathcal{X}, E) = H^n(G, E),$$

i.e., cohomology of a space with values in a sheaf and cohomology of a group with values in a G-module become cases of cohomology of a topos with values in an abelian object of the topos.

[An important factor contributing to the nontriviality of cohomology is the nontrivial contradiction between $\exists$ (image) and evaluation at X as mentioned previously. For example, in a Kripke topos in which there is a shallowest level of knowledge, $H^n = 0$ for $n > 0$. However, there are other factors, since without the shallowest level the $H^n(\mathcal{S}^{\mathbf{P}^{\text{op}}}, E)$ occur in algebraic topology and in partial differential equations under the name of "higher inverse limits."]

Note that if X and Y are any sober topological spaces (for example any Hausdorff spaces) then the topos morphisms (to be defined presently)

$$\text{Sheaves}(X, \mathscr{S}) \to \text{Sheaves}(Y, \mathscr{S})$$

are equivalent to the continuous maps $X \to Y$, and that if G and H are any groups, then the topos morphisms

$$\mathscr{S}^G \to \mathscr{S}^H$$

are equivalent to the group homomorphisms $G \to H$. Recalling that group cohomology arose in the geometric context where the group is related to the Poincaré fundamental group of a space, one sees the possible virtues of one big category Top in which spaces and groups are on "equal" footing and in which moreover a space, its universal covering space, and its fundamental group might be connected by morphisms, etc. Such a conception occurred to many people but apparently was still not sufficient to give rise to the general concept of topos; it turns out that there is a large class of topoi, the so-called étendue discussed below, that includes both these classes of examples (spaces and groups) as well as much more, and yet the "typical" topos is of still quite another kind that also arose in algebraic geometry but turned out to have still other connections with logic, apparently in particular with A. Robinson's notion of "forcing in model theory."

Some of these more typical examples arise immediately in algebraic geometry as follows. Let K be a field in $\mathscr{S}$ and let $\mathbf{A}$ be the small category of all finitely presented (finitely generated) commutative algebras over K. The category $\mathscr{S}^{\mathbf{A}}$ of all covariant functors from $\mathbf{A}$ to $\mathscr{S}$ is a topos with the following interesting property: the underlying set functor $R: \mathbf{A} \to \mathscr{S}$ is a commutative-K-algebra-object in $\mathscr{S}^{\mathbf{A}}$; if $\mathscr{X}$ is any topos defined over $\mathscr{S}$ and A is any commutative-K-algebra-object in $\mathscr{X}$, then there is a unique continuous map (geometrical morphism of topoi)

$$\mathscr{X} \xrightarrow{f} \mathscr{S}^{\mathbf{A}} \qquad \text{such that} \qquad f^*(R) \xrightarrow{\approx} A.$$

A continuous map of topoi is just a functor having a left exact adjoint f^*. Since tensor products in $\mathbf{A}$ distribute over finite cartesian products, the subcategory $\mathscr{G}$ of $\mathscr{S}^{\mathbf{A}}$ consisting of the functors that preserve finite cartesian products is also a topos, and $R \in \mathscr{G}$; the continuous maps

$$\mathscr{X} \to \mathscr{G}$$

of topoi are equivalent to the commutative-K-algebra-objects A which satisfy moreover the condition

$$a^2 = a \qquad \text{entails} \qquad a = 0 \vee a = 1$$

where the disjunction is interpreted as union of certain subobjects of A in $\mathcal{X}$. The topos $\mathcal{G}$ may also be described as the largest subtopos of $\mathcal{S}^{\mathbf{A}}$ for which the left adjoint to the inclusion functor renders isomorphic the particular inclusion map

$$\{0, 1\} \hookrightarrow A(K \times K, \_\_)$$

in $\mathcal{S}^{\mathbf{A}}$. In turn $\mathcal{G}$ has a largest subtopos $\mathcal{Z}$ for which the inclusion

$$U \cup U_{1-()} \hookrightarrow R$$

is rendered isomorphic, where for a $B \in \mathbf{A}$,

$$(U \cup U_{1-()})(B) = \{b \in B \mid \exists(1/b) \text{ or } \exists[1/(1-b)]\}.$$

The topos $\mathcal{Z}$, sometimes called the "big Zariski" topos, classifies, for any topos $\mathcal{X}$ over $\mathcal{S}$, the commutative-K-algebra-objects A in $\mathcal{X}$ that are *local* rings in $\mathcal{X}$ in the sense that $a_1 + a_2$ unit in A entails

$$(a_1 \text{ unit in } A) \vee (a_2 \text{ unit in } A)$$

where again disjunction refers to union of subobjects (of $A \times A$) in $\mathcal{X}$.

These examples represent a surprising twist of a logic that is not yet fully clarified: while a topos such as Sheaves(X, $\mathcal{S}$) for a topological space X may be "identified" as a *particular* space, a topos such as $\mathcal{G}$ or $\mathcal{Z}$ should be identified rather as a *general* concept of space in the sense that its *objects* correspond to particular spaces as follows: The Yoneda embedding, which factors through $\mathcal{Z}$, may be called "spec," since its image is equivalent to the category of affine algebraic schemes over K

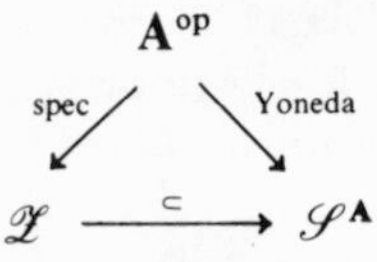

but in fact the whole category of algebraic spaces over K is fully included in $\mathcal{Z}$. For $X \in \mathcal{G}$, the morphism set $\mathcal{G}(X, R)$ is the ring of functions on the space X, and the definition of $\mathcal{G}$ may be considered as the readjustment of the notion of coproduct (from that of $\mathcal{S}^{\mathbf{A}}$) so that $\text{spec}(B_1 \times B_2) = \text{spec}(B_1) + \text{spec}(B_2)$. The latter condition is geometrically reasonable since

$$\mathcal{G}(X_1 + X_2, R) \simeq \mathcal{G}(X_1, R) \times \mathcal{G}(X_2, R)$$

holds in any case for the rings of functions, and if X_1, X_2 are determined by their global function rings B_1, B_2, it is plausible that the "disjoint" union $X_1 + X_2$ is entirely determined by its global function ring as well. Similarly, the definition of $\mathscr{Z}$ is a further readjustment of unions so that finite coverings of $X = \operatorname{spec} B$ by Zariski open subobjects (not only the clopen coverings determined by idempotents) have the correct effect on function rings. Here all Zariski open subobjects are derived from pullback from the basic one

$$\begin{array}{ccc} U & \hookrightarrow & R \\ \| & & \| \\ \mathbf{A}(K[t, t^{-1}], \_) & \hookrightarrow & A(K[t], \_) \end{array}$$

The above examples are typical in that *every* Grothendieck–Giraud topos $\mathscr{Y}$ over $\mathscr{S}$ is the classifying topos for an essentially unique theory in the way that $\mathscr{S}^{\mathbf{A}}$ is the classifying topos for the theory of commutative algebras, $\mathscr{G}$ the classifying topos for the theory of idempotentless algebras, and $\mathscr{Z}$ the classifying topos for the theory of local algebras. That is, for any topos $\mathscr{X}$ over $\mathscr{S}$, the category $\mathrm{Top}_{\mathscr{S}}(\mathscr{X}, \mathscr{Y})$ is equivalent to the category of models in $\mathscr{X}$ of the theory associated with $\mathscr{Y}$. The kind of theories that occur, which we may briefly call "positive" theories, are in general many-sorted ones, which in addition to equational axioms (such as the distributive law) may also involve axioms of the form

$$\varphi_1 \quad \text{entails} \quad \varphi_2,$$

where the φ_1 and φ_2 are formulas built up from "atomic" operations and relations using finite conjunction, possibly infinitary disjunction, and existential quantification. If only finitary disjunctions (which includes the empty disjunction "false") are involved, the topos will be a "coherent" one, and conversely. So far as models in $\mathscr{X} = \mathscr{S}$ are concerned (or more generally for any "Boolean" $\mathscr{X}$ in which classical logic reigns) any first-order theory can be construed as positive by introducing further atomic relations to play the role of any negative formulas that occur in the axioms. In this sense Deligne's theorem that every coherent topos has points is equivalent to the Gödel–Henkin completeness theorem for first-order logic and Barr's theorem that every Grothendieck–Giraud topos has sufficient Boolean-valued points implies Mansfield's Boolean-valued completeness theorem for infinitary first-order theories. However, as already pointed out, classical logic does not apply in even the simplest topological topoi $\mathscr{X}$, so that if conditions φ that essentially involve $\forall$, $\Rightarrow$ are considered, Kripke's method has to be taken into account.

As a simple example of the above, let G be a group in $\mathscr{S}$ and $\mathscr{Y} = \mathscr{S}^G$

the category of all G-sets. It is known that for any topos $\mathscr{X}$ over $\mathscr{S}$

$$\mathrm{Top}_{\mathscr{S}}(\mathscr{X}, \mathscr{S}^G) = H^1(\mathscr{X}, G)$$

up to equivalence, i.e., that continuous maps $\mathscr{X} \to \mathscr{S}^G$ correspond to principal homogeneous G-objects in $\mathscr{X}$. The latter are in fact models of the positive theory generated by one unary operation for each element g of G subject to the axioms

$(g_1 g_2)x = g_1(g_2 x)$

$g_1 x = g_2 x$ entails "false", for $g_1 \neq g_2$ (an infinite number of axioms with one free variable)

$\bigvee_{g \in G} [gx = y]$ (a formula with two free variables)

$\exists x[x = x]$

The last axiom means that for a principal homogeneous G-object X in $\mathscr{X}$, the morphism $X \to 1$ is an epimorphism in $\mathscr{X}$; it does not necessarily mean that X has a globally defined element $1 \to X$ in $\mathscr{X}$, which would imply $X \cong p^*(G_1)$ where $\mathscr{X} \xrightarrow{p} \mathscr{S}$ is the canonical grounding or "global-sections" functor and G_1 denotes G acting on itself by left translation.

Another example of central importance is the classifying topos for the theory of equality, i.e., the models of this theory in $\mathscr{X}$ are just the objects of $\mathscr{X}$,

$$\mathrm{Top}_{\mathscr{S}}(\mathscr{X}, \mathscr{Y}) \cong \mathscr{X} \qquad \text{for all } \mathscr{X}.$$

Here

$$\mathscr{Y} = \mathscr{S}^{\mathbf{S}_{\mathrm{fin}}}$$

is the category of all functors from the category of finite sets into the category of sets, with the inclusion functor U as the "generic set adjoined to $\mathscr{S}$." [If we instead consider $\mathscr{S}^{\mathbf{S}_{\mathrm{fin}}^{\mathrm{op}}}$ we get the classifying topos for the theory of Boolean algebras, with $n \rightsquigarrow 2^n$ as the "generic Boolean algebra."] Taking $\mathscr{X} = \mathscr{S}$, we see that the category of points of $\mathscr{S}^{\mathbf{S}_{\mathrm{fin}}}$ is just $\mathscr{S}$, so that in a definite sense the *objects* of any Grothendieck–Giraud topos $\mathscr{X}$ may be identified with *continuous* mappings

$$\mathscr{X} \to \mathscr{S}^{\mathbf{S}_{\mathrm{fin}}}$$

from the "space" $\mathscr{X}$ into the "space of all sets." There is a clear analogy here, to which we will return in a moment, between variable sets and variable quantities: If we replace $\mathscr{S}^{\mathbf{S}_{\mathrm{fin}}}$ by the space of real or complex numbers, we would obtain the well-known correspondence between algebras of variable quantities and spaces; however, that correspondence is perfect only for compact spaces, while for "algebras" (topoi) of variable sets the correspondence is perfect for all sober spaces and even, as we have seen, for vastly more general "spaces" as domains of variation for the variable sets. Such a "space" may be considered as the space of models for a theory, in a more refined sense than the usual one since a point determines a model, not only an elementary-equivalence class of models. It is useful to consider that the analysis of the domain of variation is a categorical refinement of the analysis in terms of parts, in the sense that any object of the topos may also be considered a generalized "part" of the domain of variation. Under the analogy the functors f^*, which are just those preserving small direct limits (addition) and finite inverse limits (multiplication), correspond to algebra homomorphisms.

We can even speak of *ideal points* of a topos in the following way. Any given infinitary theory can be construed as a finitary theory in which all formulas obtained by infinitary disjunction are reconsidered to be "atomic" formulas; in topological terms we relax the sheaf condition to consider only finite coverings. This leads to a sort of Wallman compactification

$$\mathscr{X} \subseteq \overline{\mathscr{X}}$$

for any *G–G* topos $\mathscr{X}$ (with $\overline{\mathscr{X}}$ a coherent topos), and we may consider any point $\mathscr{S} \xrightarrow{p} \overline{\mathscr{X}}$ of $\overline{\mathscr{X}}$ to be an ideal point of $\mathscr{X}$. The axiom of choice for $\mathscr{S}$ (i.e., the Deligne–Gödel–Henkin theorem) reassures us that enough ideal points exist for any $\mathscr{X}$. The above inclusion has the special properties that it generates $\overline{\mathscr{X}}$ in the sense that any $\overline{X}$ in $\overline{\mathscr{X}}$ is the canonical direct limit of all the objects X in $\mathscr{X}$ that map in $\overline{\mathscr{X}}$ to $\overline{X}$, and that the inclusion preserves *finite direct* limits (i.e., its derived functors vanish). Thus for any ideal point p of $\mathscr{X}$, the composite $\mathscr{X} \hookrightarrow \overline{\mathscr{X}} \xrightarrow{p^*} \mathscr{S}$ is a functor that preserves both finite inverse limits and finite direct limits; factoring this functor by a method due to Kock and Mikkelsen should lead to a "localization" functor $\mathscr{X} \to \mathscr{S}'$, which preserves even higher-order logic, with $\mathscr{S}'$ a two-valued topos but, in general, *not* a Grothendieck–Giraud topos (i.e., not defined over $\mathscr{S}$), making precise the idea that constancy is a limiting case of variation, but constancy is not an entirely determinate concept.

We need not always use the syntactical machinery of logic in presenting the classifying topoi $\mathscr{Y}$, since as has been elegantly utilized by Joyal and Wraith, every Grothendieck–Giraud topos over $\mathscr{Y}_0 = \mathscr{S}$ can be constructed by a finite number of applications of the following three operations:

(1) Adjoining a generic element of a given object Y_0 of a given topos

$\mathscr{Y}: \mathscr{Y}' = \mathscr{Y}/Y_0 =$ (the category whose morphisms are the commutative triangles in $\mathscr{Y}$ ending in Y_0) has the property that for any given continuous map $\mathscr{X} \xrightarrow{p} \mathscr{Y}$,

$$\mathrm{Top}_{\mathscr{Y}}(\mathscr{X}, \mathscr{Y}') \cong \mathscr{X}(1, p^* Y_0).$$

Here, if we denote by $\Pi: \mathscr{Y}' \to \mathscr{Y}$ the continuous map $\prod_{Y_0}$ whose inverse image part is just $\Pi^* = (\ ) \times Y_0$, the "generic element of Y_0" is just the element in $\mathscr{Y}'$

$$\begin{array}{ccc} 1_{\mathscr{Y}'} & \to & \Pi^*(Y_0) \\ \| & & \| \\ Y_0 & \to & Y_0 \times Y_0 \end{array}$$

determined by the diagonal map.

(2) Adjoining a generic family of objects indexed by a given object I of a given topos $\mathscr{Y}$. That is, there is a topos $\mathscr{Y}''$ with a continuous map $\mathscr{Y}'' \xrightarrow{q} \mathscr{Y}$ such that for any topos $\mathscr{X}$ and given continuous map $\mathscr{X} \xrightarrow{p} \mathscr{Y}$,

$$\mathrm{Top}_{\mathscr{Y}}(\mathscr{X}, \mathscr{Y}'') \cong \mathscr{X}/p^*(I)$$

and of course for $J \in \mathscr{X}$ we consider $\mathscr{X}/J$ as the topos of families of objects of $\mathscr{X}$ smoothly parametrized by J. In case $I = 1$, $\mathscr{Y} = \mathscr{S}$ we have $\mathscr{Y}'' = \mathscr{S}^{\mathbf{S}_{\mathrm{fin}}}$ as discussed earlier.

(3) Inverting a given morphism $Y_1 \xrightarrow{y} Y_2$ in a given topos $\mathscr{Y}$. There is a subtopos $\mathscr{Y}''' \hookrightarrow \mathscr{Y}$ (constructed with help of a Grothendieck modal operator in $\mathscr{Y}$) such that for any continuous $\mathscr{X} \xrightarrow{p} \mathscr{Y}$,

$$p \text{ factors through } \mathscr{Y}''' \quad \text{iff} \quad p^*(y) \text{ is an isomorphism in } \mathscr{X}.$$

These constructions imply some others:

(4) Adjoining a generic morphism between two given objects Y_1, Y_2 in a given topos $\mathscr{Y}$. There is $\mathscr{Y}^{\mathrm{IV}} \to \mathscr{Y}$ such that for any continuous $\mathscr{X} \xrightarrow{p} \mathscr{Y}$,

$$\mathrm{Top}_{\mathscr{Y}}(\mathscr{X}, \mathscr{Y}^{\mathrm{IV}}) \cong \mathscr{X}(p^* Y_1, p^* Y_2).$$

(5) Adjoining a generic *epi*morphism between two given objects [by applying (3) to the image of a generic morphism].

(6) Imposing equality of two given morphisms $Y_1 \rightrightarrows Y_2$ [by applying (3) to the inclusion morphism of their equalizer], etc.

In the above we have denoted by $\mathrm{Top}_{\mathscr{Y}}(\mathscr{X}, \mathscr{U})$ the category whose objects are pairs $\langle \mathscr{X} \xrightarrow{f} \mathscr{U}, \theta \rangle$, where θ is an isomorphism of functors.

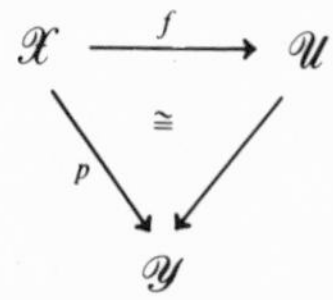

In the constructions (1) and (4) this category of continuous maps is equivalent to the small discrete category determined by the set of morphisms on the right-hand side of the condition, while in case (2) it is equivalent to the "large" (and nondiscrete) topos $\mathscr{X}/p^*I$.

For example, we may consider the functor $N:\mathbf{S}_{\text{fin}} \to \mathscr{S}$ which is constantly a countable set; it is the natural number object for the object classifier $\mathscr{Y}_1 = \mathscr{S}^{\mathbf{S}_{\text{fin}}}$ in the sense that is uniquely described as the object satisfying primitive recursion. Thus by (1) the category $\mathscr{S}^{\mathbf{S}_{\text{fin}}}/N$ has both a generic number n and a generic "set" U. The "set" of natural numbers less than n is a definite object $[n]$ of the last category, so by (5) we can further adjoin a generic epimorphism $[n] \to U$ to obtain the classifying topos $\mathscr{Y}_2$ for the theory of *explicitly finite* sets, i.e., for any $\mathscr{X} \xrightarrow{p} \mathscr{S}$, an object of the category

$$\mathscr{F}_2(\mathscr{X}) = \mathrm{Top}_{\mathscr{S}}(\mathscr{X}, \mathscr{Y}_2)$$

"is" a triple $\langle A, k, r\rangle$, where A is an object in $\mathscr{X}$, k a natural number *in the sense of* $\mathscr{X}$, and $r: [k] \twoheadrightarrow A$ an epimorphism in $\mathscr{X}$. It is easily shown that the canonical map $\mathscr{Y}_2 \xrightarrow{q} \mathscr{Y}_1$ has q^* *faithful*, so that the image of q in the sense of $\mathrm{Top}_{\mathscr{S}}$ is $\mathscr{Y}_1$ itself. But what of finite sets in themselves, i.e., those objects for which "there exists" an enumeration by some natural number?

For example, in the two-stage Kripke topos $\mathscr{S}^{\mathbf{2}^{\text{op}}}$ (i.e., the topos of sheaves on the two-point space with three open sets), there is no need for coverings in interpreting "there exists," so a finite object turns out to mean $E(X) \to E(U)$ such that both $E(X)$ and $E(U)$ are finite in the sense of $\mathscr{S}$ *and* the restriction map $E(X) \to E(U)$ itself is *surjective*. Note that a natural number $[k]$ in $\mathscr{S}^{\mathbf{2}^{\text{op}}}$ must have *identity* $[k] \to [k]$ as its restriction, so that the existence of an enumeration for E by some $[k]$ does *not* imply that E has a one-to-one enumeration by a $[k]$ (not even locally).

On the other hand if we consider $A = \{z \,|\, |z| = 1\}$, the unit circle in the complex plane, as a locally connected Hausdorff topological space, then in $\mathscr{X} = \mathrm{Sheaves}(X, \mathscr{S})$ coverings definitely do play a role; the natural number [2] is the sheaf corresponding to the étale space $X + X \to X$ over X consisting of two disjoint copies of the circle, one above the other, but if we consider the étale space $X \xrightarrow{z^2} X$ (which may be pictured as a single double loop) then the corresponding sheaf E is *locally* enumerated by (even locally isomorphic to) [2] and hence is finite.

As a third example consider $\mathscr{X} = \mathscr{S}^G$, the topos of G-sets. Here a natural number $[k]$ is a finite set with *trivial* G-action, but since there is a "covering" on which $\mathscr{S}^G$ becomes equivalent to $\mathscr{S}$ itself, a finite object is just an arbitrary finite G-set.

Our definition of finite is equivalent to the following definition, independent of the concept of natural number, which was studied by Kock, Lecouturier, and Mikkelsen. E is finite iff it is a member of the smallest subset of the power set of E, which contains 0 and all singletons, and is closed with respect to binary unions. If $\mathscr{X} \xrightarrow{f} \mathscr{Y}$ is any continuous map and Y is finite in $\mathscr{Y}$, then f^*Y is finite in $\mathscr{X}$, a property that is manifest for our definition but in fact also valid without the existence of an object N of all natural numbers in $\mathscr{Y}$. Thus taking the category $\mathscr{F}(\mathscr{X})$ of finite objects is a contravariant category-valued 2-functor of the topos $\mathscr{X}$.

Just as the enlargement

$$\mathbf{A}^{\text{op}} \subseteq \mathscr{X} \subseteq \mathscr{S}^{\mathbf{A}}$$

of the category of affine schemes was necessary because there are algebraic spaces (notably Grassman manifolds) that are not determined by a single global ring of variable quantities, so a hypothetical enlargement

$$\text{Top}_{\mathscr{S}} \subseteq \mathscr{H} \subseteq \text{CAT}^{\text{Top}^{\text{op}}}$$

of the 2-category of Grothendieck–Giraud topoi may be necessary to account for very general domains of variation that are not determined by a single global topos of variable sets. From the above discussion of finiteness, we derive the following hypothetical property of $\mathscr{H}$: A map $\mathscr{F}_2 \xrightarrow{q} \mathscr{F}$ is "epic" in $\mathscr{H}$ if for every $\mathscr{X} \in \text{Top}_{\mathscr{S}}$ and for every $E \in \mathscr{F}(\mathscr{X})$ there exist an $\mathscr{S}$-set S_i of objects of $\mathscr{X}$ with $\Sigma_i S_i \to 1$ epic and $E_i \in \mathscr{F}_2(\mathscr{X}/S_i)$ with

$$q(E_i) \cong \Pi_i^*(E) \qquad \text{in} \quad \mathscr{F}(\mathscr{X}/S_i)$$

It is not clear what the "logic," i.e., the structure of subobject lattices, for a "2-dimensional topos" such as $\mathscr{H}$ should be. Note that the lattice of *subtopoi* of a given topos is actually an *anti*-Heyting algebra; i.e., it is like the system of closed subsets (rather than open subsets) of a topological space. Thus rather than an implication operator right adjoint to conjunction, the lattice of subtopoi has a logical subtraction operator left adjoint to disjunction, so that if we define $\neg\mathscr{A}$ for a subtopos $\mathscr{A}$ to mean $1\backslash\mathscr{A}$ where 1 denotes the whole topos, then it may be useful to consider, by analogy with closed subsets of a space, that

$$\partial\mathscr{A} = \mathscr{A} \wedge \neg\mathscr{A}$$

is the "boundary" through which $\mathscr{A}$ and $\neg\mathscr{A}$ pass over into each other.

The ubiquity of variable sets suggests the following modest but rather definite conceptual guide: Endeavor to do calculations in all branches of mathematics in such a way that insofar as possible they will be valid in an arbitrary topos, not only in "the" category of constant sets, for this in many cases proves to permit a direct application of essentially naive set-theoretic techniques to higher mathematical problems in a new way that was not possible before the work of Grothendieck, Giraud, *et al.*

To pursue further the analogy with quantities, recall that many constructions that are possible for constant quantities, such as the exponential function e^x, remain meaningful and useful in any Banach algebra; similarly many constructions, notably the power set $\mathscr{P}(X) = \Omega^X$, that are possible for constant sets remain meaningful and useful in any topos. [*Caution*: $\mathscr{P}(X)$, unlike e^x, does not commute with evaluation at a point in general.] Partly independently of the fact that any commutative ring can be analyzed as consisting of local-ring-valued functions on its spectrum, it is important for many reasons to develop *linear algebra* (and hence quadratic forms, commutative algebra, Lie algebra, etc.) over an arbitrary commutative base ring. Such a development, of course, has more attendant subtleties (some of which are measured by homology) than the case when the base ring is the field of rational numbers or complex numbers. In a similar way, it is useful to develop *mathematics* over an arbitrary base topos, and this partly independently of the analysis of its objects as constant-set-valued continuous maps on some domain of variation.

Naturally there is not only an analogy but also an inclusion: Just as a system of constant quantities constitutes a (structured) constant set, so a system of variable quantities constitutes a (structured) variable set, usually with the same domain of variation. For example, if the usual construction of the complex numbers from the natural numbers is carried out in the topos of sheaves on a space, what results is just the sheaf of germs of continuous complex-valued functions on the same space. Here the "usual" construction is understood to involve defining the reals as two-sided Dedekind cuts in the rationals; one of the attendant subtleties is that one-sided Dedekind cuts constitute the sheaf of germs of *semicontinuous* real valued functions, which is in fact the natural recipient of the distance function for variable metric spaces and even of the norm for many variable C^*-algebras. Cauchy sequences in the rationals lead only to *locally constant* real-valued functions on the space.

As an example of the kind of "new direct set-theoretic approach" to a problem that is possible, consider a given continuous map $E \to X$ of topological spaces (which may be considered as a family of spaces E_x smoothly parametrized by $x \in X$). If the topos $\mathscr{X}$ = Sheaves(X, $\mathscr{S}$) is considered as the base "set theory," then as is fairly well understood E is just "one discrete set" *in* $\mathscr{X}$ if the map is étale. But what if the map is arbitrary? In any "set theory" we can, with due care (*geometrically* motivated) to the attendant subtleties of intuitionistic logic, develop general topology, sheaf theory, etc.; then (again modulo some subtleties) the induced continuous map

$$\mathscr{E} \to \mathscr{X}$$

on sheaf categories is equivalent to the global sections functor for the category $\text{Sheaves}_{\mathscr{X}}(E^{\#}, \mathscr{X})$ for a single topological space $E^{\#}$ *in* the set theory $\mathscr{X}$, with $E^{\#}$ in general not (even internally) discrete.

2.

The reader may have wondered why I mentioned my axiomatization of the category of *constant* sets as one of five developments leading to the consolidation of the concept of *variable* set. The reason is that in the development of my own thinking and in that of some of my colleagues, it was necessary to first purify the constant sets of an extraneous mental variation (along ordinals with attendant proliferation of iterated membership chains), which had been reflected in the theory of them and which is of quite a different nature from the more seriously mathematical or physical variation discussed in Section 1, in order to reveal them more starkly for what they are as well as to bring their theory closer to a reflection of actual mathematical practice. (The last was uppermost in my mind when I developed the theory ETCS for an undergraduate course in the foundations of analysis at Reed College.) Moreover, so far only the category language has successfully acted as a basis of unity for people who are trying to get clear on the general workings of variable sets (although elegant theories in the membership language have been given for the special case of topoi that can *be generated by their truth values* [subsets of 1]) and ETCS seems to have been the first categorical set theory.

There can be no doubt that in mathematical practice both sets and their membership as well as mappings and their composition play basic roles. But in setting up a formal theory one should also try to get clear on which of these is primary and which is secondary in mathematical practice.

The traditional view that membership is primary leads to a mysterious absolute distinction between x and $\{x\}$, to agonizing over whether or not the rational numbers are literally contained in the real numbers, to the "discovery" that an *ordered* pair of elements in turn *has* elements which are, however, not the original elements, and to debates over whether the members of the natural number 5 are 0, 1, 3, 4 or not, and all that is clearly just getting started; on its own formal face, a membership-based theory of sets is potentially littered with an infinite number of such formulas that even set theorists refrain from writing down due to their good mathematical sense. This situation, along with a very analogous situation with respect to the standard formalization of predicate logic, has led to the widespread view that a formalized theory and the calculations that it tries to unify are necessarily so sharply divorced from each other that only a pedant would attempt to actually use a formalized set theory, which view only helps to isolate from most people the actual advances set theorists and logicians have made.

Rejecting this pessimistic view, we may try to isolate instead the features of membership-as-primary that lead to nonmathematical set-theoretic "questions" of the kind listed. I believe the conclusion is that membership-as-primary entails membership as *global and absolute* whereas in practice membership is *local and relative*; that is, in practice we only consider membership as a relation between elements of a given set and subsets of the same given set, not between two sets given in vacuo, and the meaning of membership may vary or not vary when we transform the element and the subset according to some mapping that is, for the given context, tautological.

These considerations lead one to formulate the following "purified" concept of (constant) *abstract set* as the one actually used in naive set-theoretic practice of modern mathematics: An abstract set X has elements each of which has no internal structure whatsoever; X has no internal structure except for equality and inequality of pairs of elements, and has no external properties save its cardinality; still an abstract set is more refined (less abstract) than a cardinal number in that it does have elements while a cardinal number does not. The latter feature makes it possible for abstract sets to support the external relations known as *mappings*, which constitute the second fundamental concept of naive set theory (cardinal numbers would admit only the less refined external relations expressed by one being less than another or not). Thus "mapping" is too fundamental to be formally "defined," although we remark that a mapping satisfies the familiar $\forall x \exists ! y$ condition (and prove later that a mapping $X \to Y$ may be represented by its "graph," which is a subobject of a cartesian product $X \times Y$ as well as by its "cograph," which is a quotient object of the (disjoint) sum $X + Y$). The third concept is that of *composition* of mappings, which is defined only in case the codomain of the first mapping is the same abstract set as the domain of the second mapping (indeed otherwise the abstractness of the sets would be violated). Of course composition is *associative*, and there is an *identity* mapping for each set. Thus we have "the" category $\mathscr{S}$ of abstract sets and may speak of isomorphisms, monomorphisms, etc. It has been found possible and effective to express all structure of mathematical interest on a set (or sets) by means of given mappings.

By a "subset" we mean not a set but any monomorphic mapping, i.e., a mapping that does not permit the definition of any structure on its domain. If in the category $\mathscr{S}/Y$ of sets "over" a given set Y we restrict attention to the subsets of Y, we find that there is at most one morphism of $\mathscr{S}/Y$ between any pair of them, defining a reflexive and transitive relation $\subseteq_Y$ among the subsets of Y; thus an inclusion is something simpler and more precise than a condition such as $\forall y[y \in S_1 \Rightarrow y \in S_2]$ since it is rather a single actual mapping that respects the two subsets, although its existence will imply such a condition in view of the following definition of membership.

Suppose $T \xrightarrow{y} Y$ is any mapping and $S \rightarrowtail Y$ a subset. By abuse of notation the letter S names the subset, not only its domain. Then we define

$$y \in S$$

to mean that there exists a mapping (unique by the definition of monomorphism) rendering commutative the following diagram

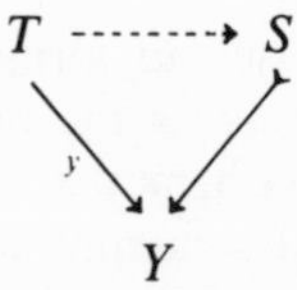

By axioms for set theory we simply mean certain properties that mathematical experience has shown to be true of sets and mappings and that are judiciously chosen so that together the axioms will imply all other properties that mathematical experience shows to be true. A fundamental axiom is the existence of one-element sets, denoted ambiguously by 1 and characterized by the property that the obvious functor

$$\mathscr{S}_{/1} \xrightarrow{\sim} \mathscr{S}$$

is an isomorphism of categories. We call a mapping $1 \xrightarrow{y} Y$ a global (or eternal) element of Y, but sometimes refer to an arbitrary $T \rightarrow Y$ as "an element of Y defined over T."

The only possible use of abstract sets T is the possibility of indexing or parametrizing things by the elements of T in the hope of clarifying actual relations between the things by means of calculations on mappings introduced to mirror the relations, and the axioms of set theory are based on the need to have abstract sets capable of parametrizing or perfectly parametrizing *mathematical* "things" that have arisen in the course of the mathematical practice of clarifying actual things and their relationships. For example, the elements of a set Y are mathematical "things" and it is precisely the mappings $T \to Y$ that may be used to parametrize these things by the elements of T; in this case perfect parametrization is possible by choosing $T = Y$ and the identity mapping.

A second kind of "thing" that we immediately need to parametrize by T is the (abstract) sets themselves; this can be accomplished by considering mappings $E \to T$ and their fibres E_t, as can be made precise with the help of additional axioms discussed below. While some such parametrizations may be half-perfect in the sense that $E_{t_1} \cong E_{t_2} \Rightarrow t_1 = t_2$, no fixed T however large can parametrize all sets, even up to isomorphism. A question that has been of much "foundational" interest, though of hardly any significance for the practice of algebra, topology, functional analysis, etc., is whether, for a given T, all imaginable families of sets parametrized by T can be represented by $E \to T$ for some E and some mapping; if "imaginable" is interpreted to mean "definable," an affirmative answer to this question is essentially equivalent (for abstract, constant sets) to the postulation of the so-called "replacement schema," whereas if $\mathcal{S}$ is considered as an object in some larger realm, an affirmative answer means that $\mathcal{S}$ itself has "inaccessible cardinality." However, in view of practice and in view of the role of $\mathcal{S}$ as a limiting case of the general notion of continuously variable sets, it seems appropriate to simply define "an internal-to-$\mathcal{S}$ T-parametrized family of objects of $\mathcal{S}$" to mean just a morphism of $\mathcal{S}$ with codomain T.

The problem of perfectly parametrizing ordered pairs of elements is solved by the axiom of cartesian products

$$\frac{T \xrightarrow{\langle y_1, y_2 \rangle} Y_1 \times Y_2}{T \xrightarrow[y_1]{} Y_1,\ T \xrightarrow[y_2]{} Y_2}$$

where projection and diagonal mappings aid in effecting the perfect parametrization and where general elements defined over arbitrary T are needed for a generally effective characterization. Note that in itself $Y_1 \times Y_2$ is just as abstract as any other set; but it has the correct cardinality and the given projection mappings enable it to be considered as having a "rectangular" structure so that its elements can be named in the usual way with help of elements of Y_1, Y_2. An important role of cartesian products is to allow consideration of algebraic operations on the elements of Y as mappings $Y \times Y \to Y$.

The immediate consideration of topoi of more variable sets is helpful even in clarifying constant sets. For example, the functor

$$\mathcal{S} \xrightarrow{T \times (\ )} \mathcal{S}/T$$

enables interpretation of a (relatively) constant set as a particular case of a "set varying over T" (that is, a family of sets parametrized by T) and in particular, identification of an element of Y defined over T in $\mathcal{S}$ with a global element of Y defined over 1_T in $\mathcal{S}/T$.

Given $X \underset{f_2}{\overset{f_1}{\rightrightarrows}} Y$ inducing a structure on X, the elements $T \xrightarrow{x} X$ for which $f_1 x = f_2 x$ can be parametrized by the domain of the *equalizer* E, which is a subset of X satisfying

$$x \in E \quad \text{iff} \quad f_1 x = f_2 x.$$

In terms of 1, $\times$, and equalizers, various constructions such as pullbacks, graphs, intersections of subsets, inverse image of a subset along a mapping, etc. can be carried out and related to each other. In particular, if $E \to T$ is considered as a family of objects parametrized by T, then the "individual" objects in the family can be extracted by means of *pullbacks*

$$\begin{array}{ccc} E_t & \longrightarrow & E \\ \downarrow & & \downarrow \\ T' & \xrightarrow[t]{} & T \end{array}$$

(in particular, consider the case $T' = 1$).

While mappings in themselves are *not* elements of any abstract set (they have, after all, "internal structure"), the mappings between two given abstract sets can be perfectly parametrized by a set of suitable cardinality with the help of an evaluation mapping. Leaving perfection aside for a moment, note that *any* mapping

$$T \times X \xrightarrow{e} Y$$

may be identified with a parametrized family of mappings $X \to Y$, since for any given $1 \xrightarrow{t} T$ we may define e_t by

$$\begin{array}{ccc} X & \xrightarrow{\quad e_t \quad} & Y \\ & {\scriptstyle \langle t_X,\, 1\rangle} \searrow \quad \nearrow {\scriptstyle e} & \\ & T \times X & \end{array}$$

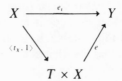

where t_X is the constant composite $X \to 1 \xrightarrow{t} T$. Then the "exponential" set Y^X, which is the right size to parametrize perfectly all the mappings $X \to Y$, is characterized by transformation rule

$$\frac{T \to Y^X}{T \times X \to Y}$$

with the evaluation $Y^X \times X \to Y$ corresponding to the case $T = Y^X$ with the identity mapping above the line. A usual slogan for interpreting the above transformation rule (often given the mysterious name "λ-conversion") says "a function of two variables is equivalent to a function-valued function of one of the variables," although of course the "name" $1 \to Y^X$ of a mapping $X \to Y$ is not "equal" to the latter but is just as abstract as any element of any set in $\mathcal{S}$. An important use of the operation Y^X is to permit consideration as mappings of functionals $Y^X \to Z$ and operators $Y^X \to B^A$; in particular, composition of mappings can itself be studied locally (i.e., for given A, B, C) as a single mapping

$$B^A \times C^B \to C^A.$$

To emphasize the basic nature of the exponentiation functor, let us consider the problem of representing the mechanical motion of matter and calculating: the distance of a particle from a fixed reference point at any given time, the motion of the center of mass of a body, and the velocity of a particle at any time. To this end, suppose M parametrizes particles in a material body (solid or fluid), E parametrizes points of space, T instants of time, and R the positive quantities. Distance may then be represented by a mapping $E \times E \xrightarrow{d} R$, which in particular gives E a convexity structure. The object E^M then parametrizes all possible (and possibly some impossible) placements of the body in space, and one of the consequences (usually calculated with the help of the theory of integration of the mass distribution intrinsic to the body) is a functional

$$E^M \to E$$

called the center of mass. If the motion of the body is represented by a mapping expressing the placement of the body at each time

$$T \to E^M$$

then this is useful (even necessary) to compute the position of the center of mass at any time by composing the two mappings. However, suppose $1 \xrightarrow{p} E$ is a fixed point and we want to calculate the distance from p to any particle at any time; then we must use the exponential transformation rule to express the same motion instead as a mapping

$$T \times M \to E,$$

which we can then compose with $E \xrightarrow{d_p} R$ to find the mapping $T \times M \to R$ to be calculated. But now note that since $T \times M \simeq M \times T$, the same motion can also be expressed as a mapping

$$M \to E^T$$

assigning to each particle the *path* it follows; indeed the motion *must* be so expressed if we are to be able to compose it with the differentiation operator

$$E^T \xrightarrow{(\,)^{\cdot}} V^T,$$

where V is the vector space of translations of E, in order to be able to transform back and thus compute

$$M \times T \to V,$$

the mapping expressing the velocity of any particle in the body at any time. I have discussed this example in some detail to emphasize the elementary character of the exponential adjointness, its necessity for science, the fact that it retains the same form for topoi other than abstract sets in which all mappings are smooth (or even more general categories), and that it must be retained in some fashion even when we consider objects that do not consist mainly of points.

Other crucial facts that follow from the existence of exponentiation, although they do not mention it, are distributive laws like

$$T \times (X_1 + X_2) \xleftarrow{\approx} T \times X_1 + T \times X_2$$

when "coproducts" (i.e., "disjoint" sums) exist, as they do in topoi. More general laws (involving pullback instead of only the special case $\times$) follow similarly from the fact (additional axiom if you wish) that the pullback functors

$$\mathcal{S}/T' \leftarrow \mathcal{S}/T$$

determined by any $T' \to T$ have *right adjoints* $\prod_t$, which fact is equivalent to the fact that each category $\mathcal{S}/T$ of T-parametrized families of sets has its own internal exponentiation satisfying the same transformation rule over T as the one already written for the case $T = 1$.

The other main use of abstract sets is to parametrize perfectly various *types of quantity*, notably positive real quantities (by R, as mentioned but not fully characterized above), and in particular matter, space, time, and the more problematical truth values (denoted by Ω in a general topos and by 2 in the case of abstract sets). We also customarily use an abstract set N to "perfectly" parametrize the finite discrete quantities. Claiming that the latter can be parametrized by an object N does not make it less problematical, since while we see (bounded pieces of) R every day, no one has yet completed a potential infinity like N. Indeed the introduction of N has given rise to all sorts of unphysical "counter examples" in analysis in the past 100 years. In this latter connection, I may remark that Professor Eilenberg's beautiful lecture on the utter simplicity of the machine needed to compute approximations to a space-filling curve need not convince one that the machine will find enough paper to complete its calculations nor that such a curve exists, since it may rather serve to clarify doubts that N exists (as opposed to 0, 1, 2, ..., each of which does of course exist). One of the reasons for referring to the quantity-type Ω as problematical is that in conjunction with reasonable properties of R it implies the existence of N.

The characterizing property of the set Ω, which perfectly parametrizes the truth values of a topos $\mathscr{X}$, is that there is a distinguished eternal truth value $1 \xrightarrow{\text{truth}} \Omega$, the inverse image $\{X \mid \varphi\}$ of which along any $X \xrightarrow{\varphi} \Omega$ is of course a subset of X for which

$$x \in \{X \mid \varphi\} \quad \text{iff} \quad \varphi x = \text{true}_T, \text{ for any } T \xrightarrow{x} X,$$

but which moreover is such that *any* subset of any X is of this form for a *unique* φ. This implies that Ω is a Heyting-algebra in $\mathscr{X}$, i.e., there is

$$\Omega \times \Omega \xrightarrow{\Rightarrow} \Omega$$

such that for $\alpha, \varphi, \psi \colon X \to \Omega$

$$\{X \mid \varphi\} \subseteq \{X \mid \alpha \Rightarrow \psi\} \quad \text{iff} \quad \{X \mid \varphi\} \cap \{X \mid \alpha\} \subseteq \{X \mid \psi\},$$

where the intersection and inclusions are as subsets of X. Still simpler is the mapping

$$\Omega \times \Omega \xrightarrow{\wedge} \Omega$$

representing intersection, which is simply the "characteristic function," in the sense of the above axiom, of the subset

$$1 \xrightarrow{\langle \text{true}, \text{true} \rangle} \Omega \times \Omega.$$

The object $P(Y) = \Omega^Y$ perfectly parametrizes the subsets of Y in the sense that the

$$T \to \Omega^Y$$

transform biuniquely into relations from T to Y (i.e., into arbitrary subsets of $T \times Y$). Indeed the foregoing sentence together with pullbacks implies the existence of $\prod$, $+$, etc., and so could be taken as the most economical axiom system for topoi (without mention yet the other quantity-types R, N, etc.). In particular, there are for each Y, mappings

$$P(Y) \times Y \xrightarrow{\ni} \Omega$$

that "locally" at Y express fully as a mapping the relation of membership between any subset of Y and any element of Y. Moreover, it follows that existential and universal quantification can themselves be expressed as mappings, e.g., by $Y^X \xrightarrow{\text{image}} \Omega^Y$ and by $PPY \xrightarrow{\cap} PY$.

The "simple recursion" property sufficient to characterize in a topos $\mathscr{S}$ a completed discrete infinity N with successor s and starting point 0 is simply the universal mapping property

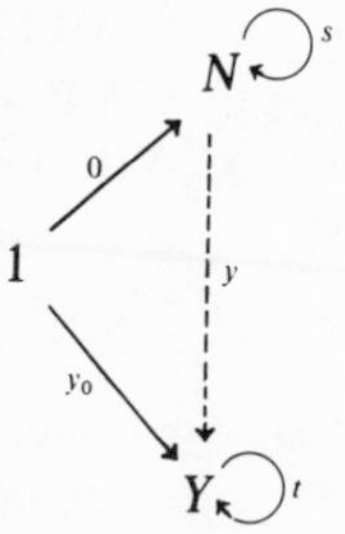

i.e., on any set Y any simple transition t and any starting point y_0 determine a unique sequence $N \xrightarrow{y} T$ satisfying two simple recursion conditions. This implies for a topos $\mathscr{S}$ that the forgetful functor

$$\text{Mon}(\mathscr{S}) \to \mathscr{S}$$

from the category of all monoids in $\mathscr{S}$ has a left adjoint W, which is the "word algebra" functor with $W(1) = N$, and a monoid homomorphism "length" $W(T) \to N$ for any set T in $\mathscr{S}$. An even stronger consequence (i.e., not equivalent in the absence of the exponention axiom true in a *topos* $\mathscr{S}$) is that for each T, the forgetful functor

$$\mathscr{S}^{(T)} \to \mathscr{S}$$

has a left adjoint that may in fact be written $W(T) \times Z \leftsquigarrow Z$. Here $\mathscr{S}^{(T)}$ denotes the category whose objects are pairs Y, θ where $T \times Y \xrightarrow{\theta} Y$ is an arbitrary $\mathscr{S}$-mapping; this category will itself be a topos provided $\mathscr{S}$ is a topos *and* N exists in $\mathscr{S}$. The adjointness expresses the expected recursion property for "generalized arithmetic" with a family T of "successors" and a family Z of "zeros." As mentioned in Section 1, a striking consequence of having N in a topos $\mathcal{S}$ is the existence of a topos $\mathcal{S}[U]$ defined over $\mathcal{S}$ such that for any topos $\mathcal{X}$ defined over $\mathcal{S}$ (i.e., equipped with a continuous map $\mathcal{X} \to \mathcal{S}$) there is an equivalence

$$\mathrm{Top}_{\mathcal{S}}(\mathcal{X}, \mathcal{S}[U]) \cong \mathcal{X}$$

of categories. It is not known at present whether, conversely, the existence of such an "object-classifying" topos over $\mathcal{S}$ implies the existence of a natural-number parametrizer N in $\mathcal{S}$.

So far we have in fact stated only axioms that in the metatheory need only the extremely weak "logic" of Descartes' analytic geometry, i.e., considering the mappings in $\mathcal{S}$ as the elements of the "universe" of discourse, we have not needed anything so powerful as so-called standard logic with its operators $\Rightarrow$, $\forall$, $\exists$, $\vee$, but rather we have expressed all the axioms only in terms of ordered pairs (triplets, etc.) of mappings subject to certain given functional operators defined on certain equationally defined "varieties" in $\mathcal{S}^2$, $\mathcal{S}^3$, etc. using in principle only substitution; yet we have expressed all the axioms (with the exception of functor "associated champ," which probably should be added) for the general theory of variable sets. This is possible chiefly because all the axioms have the form of "adjointness" except for the most tautological ones, which are purely equational (always understanding that "equational" means identities hold on certain equationally defined "subvarieties"). However, in order to express the two further axioms that in the main express how the notion of constant set contrasts with the general notion of variable set, we need to introduce into the metatheory the logical operators "there exists" and "or" in an *essential* way (I may have used these words before in this section, but always, I believe in ways that can be easily eliminated by standard methods). However, it may prove important to remark that we *still* do not use the operators $\forall$ and $\Rightarrow$ in any essential way, so that the resulting theory [equivalent to the elementary theory of the category of (constant) sets] is still a positive theory in the sense of Section 1, so that the full method of classifying topoi (i.e., generalized "spaces" of models) can in principle still be applied to this metatheory just as for, say, the theory of local rings. The two further axioms are in fact just statements claiming a close unity between the external operators "there exist" and "or" introduced into the set theory (they are written explicitly on the first page of every traditional book of formal set theory) on the one hand, and the internal operators $\exists$ and $\vee$ that exist as *mappings* in any topos $\mathcal{S}$. Explicitly, these axioms are the axiom of choice and two-valuedness:

(AC) For any $X \xrightarrow{f} Y$, if $1_Y \subseteq \exists_f(1_X)$, i.e., if f is epi, then *there exists* g such that $f \circ g = 1_Y$.

(2 val) For any $1 \xrightarrow{\varphi_1} \Omega$, $1 \xrightarrow{\varphi_2} \Omega$, if $\varphi_1 \vee \varphi_2 =$ true, then $\varphi_1 =$ true *or* $\varphi_2 =$ true, where $\vee$ denotes the mapping $\Omega \times \Omega \xrightarrow{\vee} \Omega$ representing union of two subobjects, easily defined equationally with the aid of the infinite intersection mapping $PPY \xrightarrow{\cap} PY$.

Now it is a theorem (from Diaconescu) that the axiom of choice implies the law of the excluded middle, i.e., $1 + 1 \xrightarrow{\sim} \Omega$, i.e., the subobject lattice of any object is Boolean. Thus the second axiom "2 val" (given an absolute interpretation of the external "or") expresses just that there are only two subsets of a one-element set.

It is also a theorem (easy) that the axiom of choice implies that the Boolean algebra of subsets of 1 generates the whole topos. However, although the use of "for any" and "if, then" in the statement of the above axioms is superfluous (i.e., as Gentzen sequents their structure is so simple that the whole system of rules of inference for the metatheory could easily be set up to eliminate these phrases entirely), to state, on the other hand, for a general (not necessarily Boolean) topos what "generating" means (for a class of objects, not necessarily a class of subobjects of 1) requires an *essential* use of $\forall$, $\Rightarrow$ in the metatheory, so that classifying topoi in themselves will reveal less about such a theory. Namely, if $\mathscr{A} \subset \mathscr{E}$ is a given class of objects in a topos $\mathscr{E}$, then "$\mathscr{A}$ generates $\mathscr{E}$" just means that the law of extensionality holds in $\mathscr{E}$ for elements defined on objects A in $\mathscr{A}$, i.e.,

($\mathscr{A}$ gen) For any two subsets S_1, S_2 of any Y, if *for every* A in $\mathscr{A}$ and *every* $A \xrightarrow{y} Y$, $y \in S_1$ *implies* $y \in S_2$, then $S_1 \subseteq_Y S_2$.

3.

In this part all topoi will be over a fixed base $\mathscr{S}$, which we may assume is constant sets to make some things easier to state. Remember, meanwhile, that the basic philosophy is to try to allow $\mathscr{S}$ to be as general as possible in the hope of applying results directly to the case where $\mathscr{S}$ is (the set-valued sheaves on) a topological space, or is (the permutation-representations of) a group, or indeed where $\mathscr{S}$ itself is "a category of spaces" whose points are typical algebras in which "functions" on the "spaces" have their values. Recall that a continuous map $\mathscr{X} \to \mathscr{Y}$ means a functor having a left exact left adjoint and that "$\mathscr{X}$ has a set of generators" means that there is an object in $\mathscr{S}$ that parametrizes a family of objects of $\mathscr{X}$, which in turn satisfies (some internal version of) the extensionality condition stated for an unparametrized class $\mathscr{A}$ at the end of Section 2.

Suppose $Y_{\mathscr{X}}$ is an object of $\mathscr{X}$; what could it mean in terms of the topoi involved that $Y_{\mathscr{X}}$ is actually the sheaf of germs of continuous maps $\mathscr{X} \to \mathscr{Y}$, imagining for the moment that $\mathscr{Y}$ "is" a topological space? A reasonable definition would seem to be the following: for any object U of $\mathscr{X}$, there is a natural equivalence of categories

$$\mathrm{Top}_{\mathscr{S}}(\mathscr{X}/U, \mathscr{Y}) \cong \mathscr{X}(U, Y_{\mathscr{X}}).$$

Here we recall that if $\mathscr{X}$ "is" also a space and if it happens that $U \subseteq 1$, then $\mathscr{X}/U$ "is" just U considered as a space in its own right, while $\mathscr{X}(U, Y_{\mathscr{X}})$ is the set of sections over U of the sheaf $Y_{\mathscr{X}}$. The fact that the left-hand side of the equation above is a category, usually not discrete, makes us realize that we should have taken $Y_{\mathscr{X}} \in \mathrm{Cat}(\mathscr{X})$ as a category object in $\mathscr{X}$ and not just a plain object in general; this is necessary even if $\mathscr{Y}$ is a non-Hausdorff sober space, since relations of the kind $y' \in \overline{\{y\}}$ give rise to (in that case unique) morphisms $y' \to y$ of points. Thus we in general expect $Y_{\mathscr{X}} \in \mathrm{Cat}(\mathscr{X})$.

Now we put the question: Which $\mathscr{Y}$ will have a sheaf of germs of continuous functions $\mathscr{X} \to \mathscr{Y}$ for all $\mathscr{X}$ or for all $\mathscr{X}$ having a set of generators? It is clear that for most $\mathscr{Y}$ this will not be the case since $\mathscr{X}(U, Y_{\mathscr{X}})$ has to be a *set* (parametrizable by an object of $\mathscr{S}$) whereas for $\mathscr{Y}$ = the Zariski topos, or most any classifying topos, there is a proper class of models for the associated theory, and hence $\mathrm{Top}_{\mathscr{S}}(\mathscr{X}, \mathscr{Y})$ is not equivalent to a set, even for $\mathscr{X} = \mathscr{S}$ ("1 point space $\mathscr{X}$"). On the other hand, if $\mathscr{Y}$ is a topological space but $\mathscr{X}$ *arbitrary*, it is easily seen that

$$\mathrm{Top}_{\mathscr{S}}(\mathscr{X}, \mathscr{Y}) \subseteq \mathscr{S}(\mathscr{Y}(1, \Omega_{\mathscr{Y}}), \mathscr{X}(1, \Omega_{\mathscr{X}})),$$

and with slightly more effort one sees that $Y_{\mathscr{X}}$ exists. Indeed since $\mathscr{S}$-valued points of $\mathscr{Y}$ play no particular role, it is irrelevant whether $\mathscr{Y}$ has enough points, so we may assume simply that $\mathscr{Y}$ is generated by its subobjects of 1 [i.e., $\mathscr{Y}$ is the canonical sheaves on the complete Heyting algebra $\mathscr{Y}(1, \Omega_{\mathscr{Y}})$ in $\mathscr{S}$] and get the same result. But there are still many more $\mathscr{Y}$ for which $Y_{\mathscr{X}}$ "always" exist.

Dropping the significant but entirely distinct question of sufficiency of points, we generalize Grothendieck's definition of *étendue* to obtain the definition we will use:

Definition. $\mathscr{Y}$ is an étendue over $\mathscr{S}$ iff there exists $C \twoheadrightarrow 1$ (epic) in $\mathscr{Y}$ such that $\mathscr{Y}/C$ is generated (over $\mathscr{S}$) by its subobjects of 1.

Briefly, the slogan is that $\mathscr{Y}$ is "locally a topological space." In general, to define a property as holding locally one has to allow the covering C to be of the form $C = \sum C_i$ and ask for the property on each $\mathscr{Y}/C_i$; however the property of being a topological space is "additive" so we can use the simpler notion of covering. The first example of an étendue seems to have been the space of moduli of algebraic curves, which is prevented from being globally a space due to the action of Galois groups within each point. Yes, something vaguely reminiscent of particle spin is going on in such spaces, and the most naked form is that for any group G, the category $\mathcal{S}^G$ of G-sets is an étendue with only one point! This is easily seen from the observations that $\mathcal{S}^G/G \cong \mathcal{S}$ and that $G \twoheadrightarrow 1$ where the last two G's denote the regular representation.

A general explanation of why étendues arise in topology is that the "inclusion" functor

$$\mathrm{top}(\mathcal{S}) \xrightarrow{\mathrm{sh}} \mathrm{Top}_{\mathcal{S}}$$

does not in general preserve coequalizers; in particular, suppose a group G acts on a space X and consider the coequalizer diagram for the notion of orbit space—then if the action is good in some recognized sense, $\mathrm{sh}(X/\!/G)$ will also be the (2-) coequalizer in $\mathrm{Top}_{\mathcal{S}}$, while if the action is bad the latter coequalizer tends to be an étendue that is *not* a space. (It seems to me that functional analysis *internally* in such a topos may shed light on hard cases in harmonic analysis and ergodic theory, but I have not had time to investigate this in detail.)

Actually the group case is not quite typical, contrary to what is suggested by some exercises in SGA4; if $\mathcal{Y}$ is any topos having a set $\mathcal{A}$ of generators such that for $A, A' \in \mathcal{A}$, $\mathcal{Y}(A, A')$ consists entirely of *mono*morphisms, then $\mathcal{Y}$ is an étendue, since $C = \sum_{A \in \mathcal{A}} A$ works. For example, consider the topos $\mathcal{S}^{\circlearrowleft}$ whose objects are just sets each equipped with an arbitrary endomorphism, for which $\langle N, s\rangle$ is a convenient generator; then

$$\mathcal{S}^{\circlearrowleft}/\langle N, s\rangle \cong \mathcal{S}^{\omega}$$

where ω denotes the ordered set of natural numbers; the topos $\mathcal{S}^{\omega}$ even has enough points to be a topological space—these points are just the natural numbers plus one more point at infinity whose stalk functor is

$$\mathcal{S} \xleftarrow{\lim} \mathcal{S}^{\omega}.$$

This shows that $\mathcal{S}^{\circlearrowleft}$ is an étendue, which in fact has exactly two points $\{\mathbb{Z}\}$ = (the image under the projection of the point at ∞) and $\{N\}$ = (the image under the projection of each and every finite point n of the "covering" space); not all endomorphisms in the category of points are isomorphisms though.

Now it is easy to make the conjecture that the only $\mathcal{Y}$ that "always" have sheaves $Y_{\mathfrak{X}}$ of germs are the étendue. If so, study of both classes could perhaps be deepened. We have seen that many "domains of variation" are too big to be sets (but it would be even worse to consider them as "abstract classes," for example, since part of their essence is a very strong 2-topological character expressed by the concept of a topos or something like it). On the other hand perhaps, at least the idea that every *type of quantity* is a set could be maintained ("arbitrary cardinals" as a "type of quantity" would actually be quite far from mathematical practice) then the old ideas of Descartes, and more explicitly Riemann, that "every" domain $\mathcal{Y}$ of variation is isomorphic to a part of a type of quantity could be retained simply as the definition of a particular kind ("quantitative" as a special case of "qualitative") of domain; but such a definition in the present context would seem to reduce to our condition that $Y_{\mathcal{X}}$ exist for all $\mathcal{X}$. The idea that, for example, a cohomology class $\mathcal{X} \xrightarrow{\alpha} \mathcal{S}^G$ is a sort of variable quantity of type G varying over $\mathcal{X}$ has a definite intuitive appeal, in spite of the fact that α vanishes at every *point* of $\mathcal{X}$.

REFERENCES

[1] Lawvere, F. W. Continuously variable sets: algebraic geometry = geometric logic, *Proc. Logic Colloq. Bristol, 1973.* North-Holland Publ., Amsterdam (1975), 135–156 (and the references cited therein).

[2] Lawvere, F. W., Maurer, C., and Wraith, G. C. "Model Theory and Topoi" (Springer Lecture Notes No. 445). Springer-Verlag, Berlin and New York, 1975.

Research partially supported by Consiglio Nazionale de Ricerche d'Italia and the University of Chicago.

AMS 18B05

DEPARTMENT OF MATHEMATICS
STATE UNIVERSITY OF NEW YORK AT BUFFALO
AMHERST, NEW YORK

The Work of Samuel Eilenberg in Topology

SAUNDERS MAC LANE

The time is April 25, 1939. The world is uneasy. Just last fall, Chamberlain had made the Munich agreements with Hitler, and by now it is clear that they have not established Peace in our time. There is fear in the air. This is reflected in the mathematical community. Stimulated by the growth of mathematical research, and supported by Otto Neugebauer's knowledge of the workings of the *Zentralblatt für Mathematik*, the American Mathematical Society is drawing up plans to publish its own reviewing journal. The Germans are concerned by the prospect, so the Springer-Verlag has sent an emissary, F. K. Schmidt, to negotiate with the Council of the AMS. He has come to an AMS meeting in Cambridge, Massachusetts—and incidentally, to talk with Saunders Mac Lane about a joint paper they might prepare to rectify some errors in an earlier paper by Schmidt on the generation of inseparable extensions of fields of prime characteristic.

Algebraic topology is growing and solving problems, but non-topologists are very skeptical. At Harvard, Tucker (or perhaps Steenrod) gave an expert lecture on cell complexes and their homology, after which one distinguished member of the audience was heard to remark that this subject had reached such algebraic complication that it was not likely to go any further. The textbook presentations of topology were sparse; there was

Hausdorff on Mengenlehre and Kerekjarto on Topology. Veblen's *Analysis Situs* lacked certain geometric insights, and Lefschetz's 1930 *Topology* lacked some algebraic precision. That excellent text by Seifert and Threlfall had appeared, while point set topology had been codified in permanent fashion (*pace* Bourbaki and the current crop of categorical topologists) in the first of three projected volumes of Alexandroff-Hopf. That first volume treated only homology groups, but cohomology groups had just been recognized. Indeed, at the 1936 topology conference in Moscow, three people had independently formulated the Alexander–Whitney formulas for the cup product of cochains. Hopf had described the essential maps of S^{n+1} on S^n, Hurewicz had proved that higher homotopy groups mattered, and Whitney had shown that the Hopf classification theorem (for maps of an n-complex into an n-sphere) could be stated better by using cohomology groups. In Poland, Kuratowski led a very vigorous school, with notable results flowing from his students: Borsuk on retracts and Eilenberg on the topology of the plane. In the United States, point-setters vied with combinatorial topologists. Topology was on the move.

April 25, 1939 is a significant date in this movement: On that date, Sammy arrived in the United States. He was not yet here in person—that was to be two days later, after his brief stop in England to see Henry Whitehead. But on this date he first submitted a paper to the *Annals of Mathematics*. This paper, "Cohomology and Continuous Mappings," was a natural continuation of an earlier Eilenberg opus (*Fundamenta*, 1938) where he had first used cochains with coefficients in a homotopy group. The *Annals* paper was indeed a careful formulation of a particular case of obstruction theory. Better known is his slightly later paper on the "Extension and Classification of Continuous Mappings" given at the 1940 Conference on Topology at the University of Michigan. In that paper, Sammy formulated in explicit, clear, and easily usable form all the essential properties of obstructions to the extension of continuous maps, codifying effective but vague ideas then "in the air" (Whitney, *et al.*). Ever since that time, this formulation has been the standard for all obstruction theory in all its varied uses, including even the recent Wall obstructions to surgery.

This paper is a fine example of Sammy's influence and style—the same style is also exhibited in his 1944 paper on "Singular Homology," in which he carved out (from confusing discussions in various texts about oriented simplices and changes of signs) the clear and direct definition of the singular complex. This is the definition which we still use.

That 1940 paper on obstructions also was a start for a good deal of Sammy's own subsequent work. Elements of Ext, cohomology classes for groups, and k-invariants for spaces (and for Postnikov towers) are all examples of obstructions, while an Eilenberg–Mac Lane space is one of the spaces in which these obstructions live. Sammy both clarified obstructions and followed up their nature and habitat.

With his actual arrival in this country, Sammy plunged at once into America (via hitchhiking) and into American mathematics. The group of active mathematicians was small then, and Sammy soon knew them all. He received vital encouragement (as had many other youngsters) from Solomon:

Here's to Lefschetz, Solomon L.
Irrepressible as hell.
When he's at last beneath the sod
He'll then begin to heckle God.

There was more encouragement from Ray Wilder, where Sammy had his first U.S. position at the University of Michigan. There he rapidly displayed his exceptional talent for collaboration: joint papers with E. W. Miller, O. G. Harrold, R. L. Wilder, Deane Montgomery and others, as well as later collaboration with Tudor Ganea, H. Cartan, and John Moore—witness the beautiful and decisive 1965 joint paper in which appeared the notion of algebra for a monad (there misnamed "triple") an idea which dominated category theory for five or six years. And note especially the collaboration with Steenrod on the *Axiomatic Approach to Homology Theory*, which led to the book *Foundations of Algebraic Topology*. This book (even though Volume II never did appear) effectively cured the previous troubles about the lack of a text on topology. Moreover, Sammy advertised its beauties so well that its early version had an enormous influence long before it actually appeared in print.

As of collaboration, I'd like to report on the birth of Eilenberg–Mac Lane. During the period 1938–1941, I had collaborated extensively with the late O. F. C. Schilling in work on division algebras over local fields and their description by factor sets (two-dimensional cocycles of a group). In fact, Schilling and I, working in nonabelian class field theory, had sharply noticed the lack of any three-dimensional cocycles. These uses of two-dimensional factor sets had led me to consider the parallel use of factor sets to describe group extensions. These extensions I had studied considerably, especially the extensions of abelian groups, and I had given an hour talk on this subject at an AMS meeting in New Orleans. Then in the spring of 1941 T. H. Hildebrandt invited me to give a series of six Alexander Ziwet lectures at the University of Michigan. In these lectures, I talked about group extensions; Sammy listened carefully, but he had to leave after my fifth lecture, so he asked me to give him a private summary of the sixth. I did. For this lecture I had calculated what seemed to me a significant example of the group of abelian group extensions [in present notation, the group Ext (G, H)] for the case when G is the additive group of p-adic integers. After I explained these calculations, Sammy remarked that almost the same problems came up in a recent paper by Steenrod on "Regular Cycles of Compact Metric Spaces"; Sammy suggested that my methods might help settle a question about the p-adic solenoid which Steenrod had not resolved. At first I didn't see how, but after an all-night session I did: More results must flow from this, beyond the fact that the p-adic solenoid was the dual of the p-adic integers. We resolved together to get to the bottom of this curious connection between algebra and topology. It took us 14 years (1941–1955) and 15 papers, not counting research announcements. We viewed these papers as musical compositions numbered from Opus I to Opus XV. For many years Sammy was fond of saying that 14 of them were still alive—then and perhaps even now.

Opus I, "Group Extensions and Homology" formulated the universal coefficient theorem for cohomology and settled Steenrod's problem. This theorem had to use Ext(A, B) for abelian groups, and depended essentially upon the calculation of this group Ext by means of a free resolution of A. Since it was just a "short" resolution, we didn't recognize it as such or use modules, to say nothing of projective modules, and we did carefully include a proof that what we now would call $\mathrm{Ext}^2(A, B)$ was zero—so we knew the subject stopped there, with Ext^1.

In a different direction, questions were open—very much so. We treated universal coefficient theorems for Čech cohomology, so we had to handle limits of inverse systems of groups. To construct isomorphisms between such limits, we needed transformations between two inverse systems; moreover, we knew that the universal coefficient theorem for complexes was "natural" and we wanted to make this statement a real *theorem* and not a pious platitude. To do both these things we had to discover the notion of a natural transformation. That in its turn forced us to look at functors, which in turn made us look at categories. These were very general notions indeed. A crucial step was then our willingness to write a paper on these generalities— Opus II, "General Theory of Natural Equivalences," published in the *Transactions of the AMS*, 1945.

For many years (until about 1958) this was the only paper in its subject; it still is a good introduction. Moreover, Sammy has played a role in almost all the decisive steps that have subsequently expanded category theory in the following stages.

Abelian Categories first appeared in my 1950 paper on "Duality for Groups"; this study arose from discussion with Sammy and my wish to do Eilenberg–Steenrod axiomatic homology theory with values not just abelian groups but objects of a more general abelian category. The next step toward abelian categories was taken independently by A. Grothendieck in Kansas and by D. Buchsbaum in a Columbia thesis suggested by Eilenberg.

Adjoint Functors. The decisive paper (1958) is by Dan Kan, influenced by Eilenberg. With this idea, the subject required a book. The first books on *Abelian Categories* (Peter Freyd) and *Theory of Categories* (Barry Mitchell) were started when the authors were at Columbia with Sammy.

Algebraic Theories and other vital topics in Lawvere's 1963 thesis marked a turning point in category theory and Lawvere, though thoroughly independent, was a Ph.D. student of Sammy's.

Triples, the next big topic, was started by the 1965 Eilenberg–Moore paper.

Closed Categories and the related ideas of categories relative to a closed category, were (too voluminously) codified in the big Eilenberg–Kelly paper in 1966.

Elementary Topoi is the next (and last until now) major step. Sammy had no direct hand in this, but the major contributors in this country were two of Sammy's former students: Lawvere and Tierney.

Thus today when our hero is wont to give lectures (at the First International Symposium on Category Theory Applied to Computation and Control) about the use and misuse of categories, we must still recognize that fine Eilenberg touch all along the line of development.

I return to the work of the Eilenberg–Mac Lane firm. The next major impetus was a beautiful 1942 paper by H. Hopf, "Fundamentalgruppe und Zweite Bettische Gruppe," which showed that a certain part of the second homology group of a space depended (functorially) on the fundamental group. Hopf's paper did not spell out the algebraic form of the dependence, but a rapid study of his results convinced us that this was the sort of dependence which we knew how to understand, algebraically, and we set out to do this. We were right; our techniques would apply. We were led to the main theorem of our Opus III, asserting for an aspherical space X that the cohomology groups $H^q(X, G)$ are exactly the cohomology groups of the fundamental group $\Pi_1(X)$. Indeed, to formulate this theorem we were forced to give the first complete formulation of the cohomology of groups. H. Cartan, in his lectures at this symposium, has discussed some of the consequences of the discovery of the cohomology of groups, such as the effects on homological algebra and the description of the cohomology of other sorts of algebraic systems. At the time, we were also keenly aware of the connection with algebraic number theory and class field theory, and started work in this direction in Opus VII, on "Cohomology and Galois

Theory," but did not pursue it further. Mathematical life is full of choices; we were aware that much more lay in this direction. Emil Artin (about 1946) urged us to continue; we didn't. Mathematics has many branchings, more than one of them fruitful, and it takes many active people (in this case, Hochschild and Tate) to follow them up.

At this point, the firm of Eilenberg–Mac Lane should perhaps record its considerable debt to the NDRC (National Defense Research Council). The time is July 1944 and the world is caught up with an activity that threatens to overwhelm or postpone any thoughts about categories, aspherical spaces, cohomology of groups, or any combination thereof. Mac Lane, *faut de mieux*, finds himself in New York as Director of the Applied Mathematics Group of Columbia University, instructed to hire many fresh mathematical brains to help with the research side of the war effort. One of his first acts was to hire Samuel Eilenberg—as well as Irving Kaplansky, George Mackey, Donald Ling, and many others. During the day we all worked hard at airborne fire control, after which Sammy and I would go out for dinner, followed by an evening of work on categories and/or cohomology. At this remove in time, I can't be sure that the work on fire control (despite our devotion) had any real effect; war research involved many difficulties, some real and some bureaucratic; these were duly noted in my final report for the Applied Mathematics Group. That report (more exactly, part 2 on "Aerial Gunnery Problems," as cited in the bibliography) was initially classified *confidential* and hence buried in the Government Archives. By now it is declassified, but hardly interesting. Fortunately, the same constraints did not apply to mathematics. The record in categories and cohomology is there for the world to judge—in particular the fact that, thanks to inadequate wartime communication, the cohomology of groups was discovered independently by Beno Eckmann and the homology of groups independently both by Hopf and by Freudenthal.

I return to the work of the firm. The next big project, started by Opus IV, was the study of Eilenberg–Mac Lane spaces $K(\Pi, n)$—those spaces (or *CW* complexes) whose only homotopy group is the abelian group Π in dimension n. Given the earlier theorem relating the aspherical spaces [the spaces $K(\Pi, 1)$] and the cohomology of groups, these spaces $K(\Pi, n)$ seemed to us to be naturally there, and Hurewicz had already proved that the cohomology of such a space did depend only on Π and n, exactly as in the case $n = 1$. Except for projective spaces, there are not many examples of such spaces which occur in nature. I can recall very clearly my first lecture about them in a seminar (about 1946) at Harvard. Hurewicz was there—and expressed a well-founded skepticism about their real existence. Nevertheless, it was clear to us that these Eilenberg–Mac Lane spaces could be the building blocks for more general spaces. This idea appeared explicitly in our Opus VIII (and in Opus IX), where the first k-invariant of space (one with two homotopy groups) was defined, and where we showed that in principle the cohomology of such a space was a function of the two homotopy groups and the k-invariant. The proof depended on Eilenberg's understanding (again presented in his paper on singular homology) of the construction of a *minimal* subcomplex of the singular complex of a space. In 1947 we also had an example of two spaces with the same Π_1 and Π_2 where a simple change in the k-invariant made a difference in the homotopy type of the space. Also in 1947 both Eilenberg and (independently) Zilber had an understanding of what is today called the Postnikov system of a space. Indeed, the idea of such a system (or "tower") is a direct generalization of the Eilenberg–Mac Lane notion of a k-invariant—although at that time, in 1947, it would indeed have been pretty complicated to express the notion explicitly in its full generality.

The Eilenberg–Mac Lane firm turned its attention in Opus X and Opus XII to a more explicit problem: Since the homology and cohomology of a space $K(\Pi, n)$ depend (algebraically) on Π and on n, and since this dependence for $n = 1$ is given by the homology and cohomology of the group Π, generalize by finding an explicit form for this dependence for all n. From conversations with George Whitehead we also knew that better information about the homology of $K(\Pi, n)$ could be exploited to yield good information on the homotopy groups of spheres. Early in the fall of 1947 we convinced ourselves that it would be possible, with a little calculation, to find out about the dependence, so we set about to do so. It was possible, but it took us seven years and mountains of calculations, carefully paginated in many indexed series of papers and kept orderly in many notebooks. My only regret is that I didn't keep that pile permanently in my office as an example to the young. Never before or since have I worked so hard and long on a single project. On only one other occasion have I had to handle and sort so much paper—that other occasion was recent, in connection with the review for the National Academy of Sciences of the controversial report on the effect of herbicides in Vietnam.

To start our analysis of the complex $K(\Pi, n)$ we had an exact description of the complex, but a very clumsy one, as one may see by comparing the formulas in Opus III and Opus IX with any current formulation, in which the cells of $K(\Pi, n)$ are described as n-cocycles on Δ^q. We could see at once that our description of $K(\Pi, n)$ was not only clumsy, but needlessly big—for low n and low dimension we could replace this model of $K(\Pi, n)$ by much simpler chain complexes depending on Π and

n. So replace it we did, in case after case, with extensive comparisons by chain equivalences and various attempts at geometric and simplicial pictures that never quite came into focus. What did gradually appear was that $K(\Pi, 1)$ for Π abelian had a product—the shuffle product—and that $K(\Pi, 2)$ could be described in terms of $K(\Pi, 1)$ and the shuffle product by cells, which we wrote as $[\sigma_1|\cdots|\sigma_k]$, where the various σ_i, separated here by bars, are cells of our well-known complex $K(\Pi, 1)$, and where the boundary is given in terms of the shuffle products and the boundary in $K(\Pi, 1)$. Once we had this notation under good computational control, it was natural to call this process—the passage from $K(\Pi, 1)$ to an equivalent of $K(\Pi, 2)$ or from $K(\Pi, 2)$ to $K(\Pi, 3)$—the "bar" construction. And so, in the fullness of time, we did prove (in Opus XII) that $K(\Pi, n)$ is chain equivalent to the complex obtained by applying n bar constructions to the integral group ring of Π:

$$K(\Pi, n) \simeq B^n(Z(\Pi)).$$

Only after the fact did we realize that our symbol "bar" was really a tensor product and that the shuffle products made each $K(\Pi, n)$ into what we now call a differential graded ring. But even without that, with the basic result in hand, we were able to go on to get invariant formulas in closed form for certain homology groups of $K(\Pi, n)$. For example, $H_4(\Pi, 2)$ is the universal quadratic functor $\Gamma(\Pi)$ of the group Π.

Indeed, these homology groups appear to be some of the most relevant examples of nonadditive functors of groups. There is a considerable range in which $H_{n+k}(\Pi, n)$ is quadratic in Π and then further ranges where it is cubic, quartic, etc., all according to the concepts of higher degree functors as explained in Opus XIII. However, some of these functors were difficult to compute by the methods we then used; for those methods the critical case was $H_7(\Pi, 2)$. As Opus XIII states, this involved a full use of the Künneth formula and so was not worked out in that paper.

Shortly after the publication of Opus XIII there appeared the masterful Cartan seminar (1954–1955) on homotopy and Eilenberg–Mac Lane spaces. This seminar made more use of the conceptual properties of the bar construction and so yielded a complete calculation of the homology of an Eilenberg–Mac Lane space with coefficients integers modulo p. Going further, the Cartan seminar provided a way of giving the homology with integer coefficients. The description was to use the homology operations applicable for coefficients mod p and construct from them the tensor product of a number of elementary complexes, one for each appropriate homology operation. The homology of this tensor product reduced by certain relations then provided the required integral homology of the Eilenberg–Mac Lane space. This gave an explicit calculation, by not quite an invariant functorial formula in the sense of our earlier Opus XIII.

Today, I am happy to report that these matters have been pushed further, so that now the homology formulas for $H_*(\Pi, n)$ are all in hand. The thesis of Ross Hamsher considered the homology of an Eilenberg–Mac Lane space $K(\Pi, 1)$ for Π abelian and gives complete invariant formulas for the homology of such a space. The next step has been carried out in the thesis of Gerald J. Decker, Chicago, 1974, which obtains in principle invariant functorial formulas for all the homology groups of an Eilenberg–Mac Lane space—including in particular the critical group $H_7(\Pi, 2)$. This is done by combining Cartan's results from the 1954–1955 seminar with a new version of the Künneth formula. The new version applies to a special class of complexes, those in which the Bockstein map has a left inverse. For these complexes one can replace the usual Künneth formula

$$0 \to \sum_{p+q=n} H_p K \otimes H_q \mathrm{L} \to H_n(K \otimes \mathrm{L}) \to \sum_{p+q=n-1} \mathrm{Tor}(H_p K, H_q \mathrm{L}) \to 0,$$

where the exact sequence is split but not naturally split, by a single closed formula. This closed formula then applies to the tensor product of elementary complexes used in the Cartan seminar as above, and from this Decker derives the complete homology formula.

It is thus a pleasure to honor Eilenberg's 60th Birthday with a report that one of the outstanding questions in his work has now been settled. There of course remain a number of other problems that haven't yet been settled. For example, right in the same line, one would wish a full understanding of the cohomology of Eilenberg–Mac Lane spaces in terms of suitable invariant formulas. The Eilenberg–Mac Lane Opus XIII achieved some results in this direction and Decker's thesis gives more, but we are still far from a complete description. Moreover, the case of Eilenberg–Mac Lane spaces is, as always, just a first step. There must be some further way of conceptually understanding how the homology and cohomology of a Postnikov tower depends on the homotopy groups and the k-invariants.

Another long-standing question is that of the interpretation of the higher homology groups of a group. H^2 has a clear interpretation by group extensions, while H^3 has an interpretation by obstructions to the existence of an extension—as presented in our Opus V, which was a sequel to the earlier work of Reinhold Baer. This Opus V has had extensive influence because the same sort of interpretation can be made (and has been made) for many other sorts of cohomology theories. There does remain, however, a question whether one could find a corresponding meaning for the higher cohomology groups. The Eilenberg–Mac Lane paper on loops did find such an interpretation, but not one that was terribly successful. However, recent work by John Duskin offers such interpretations by the use of his concept of a Π-torsor.

Generic acyclicity, finally, is one open problem still much on our minds. The idea was broached in Opus XV, where we observed that most of the typical cohomology theories were described in terms of chain complexes that are very nearly acyclic—"generically" acyclic in a precise sense described there. The further reaches of this observation leave open questions that are still mysterious and fascinating. Let us hope that at the 70th Birthday celebration we can report on these.

BIBLIOGRAPHY

Baer, R., Erweiterung von Gruppen und Ihren Isomorphismen, *Math. Z.* **38** (1934), 375–416.

Cartan, H., and Eilenberg, S., "Homological Algebra." Princeton Univ. Press, Princeton, New Jersey, 1956.

Cartan, H., Sur les groupes d'Eilenberg–Mac Lane $H(\Pi, n)$, I and II, *Proc. Nat. Acad. Sci. U.S.A.* **40** (1954), 467–471 and 704–707.

Cartan, H., Algèbres d'Eilenberg–Mac Lane et homotopie, Séminaire (Ecole Normal Sup.), Paris, 1955.

Decker, G. J., The integral homology algebra of an Eilenberg–Mac Lane space. Ph.D. Thesis, University of Chicago, Chicago, 1974.

Duskin, J., $K(\Pi, n)$-torsors and the interpretation of "triple" cohomology, *Proc. Nat. Acad. Sci. U.S.A.* **71** (1974), 2554–2557.

Eckmann, B., Der Cohomologie-Ring einer beliebigen Gruppe, *Comm. Math. Helv.* **18** (1946), 232–282.

Eilenberg, S., Transformations continues en circonference et la topologie du plan, *Fund. Math.* **XXVI** (1936), 62–112.

Eilenberg, S., Sur le prolongement des transformations en surfaces sphériques, *Fund. Math.* **XXXI** (1938), 179–200.

Eilenberg, S., On the relation between the fundamental group of a space and the higher homotopy groups, *Fund. Math.* **XXXII** (1939), 167–175.

Eilenberg, S., Cohomology and continuous mappings, *Ann. of Math.* **41** (1940), 231–251.

Eilenberg, S., Extension and classification of continuous mappings, *Lectures in Topology, University of Michigan Conference of 1940*, Univ. of Michigan Press, Ann Arbor, 1941.

Eilenberg, S., On spherical cycles, *Bull. Amer. Math. Soc.* **47** (1941), 432–434.

Eilenberg, S., Singular homology theory, *Ann. of Math.* **45** (1944), 407–447.

Eilenberg, S., Homology of spaces with operators, *Trans. Amer. Math. Soc.* **61** (1947), 378–417.

Eilenberg, S., and Ganea, T., On the Lusternik–Schnirelmann category of abstract groups, *Ann. of Math.* **65** (1957), 517–518.

Eilenberg, S., and Harrold, O. G., Jr., Continua of finite linear measure **I**, *Amer. J. Math.* **65** (1943), 137–146.

Eilenberg, S., and Kelly, G. M., Closed categories, *Proc. Conference on Categorical Algebra, La Jolla, 1965* pp. 421–562. Springer-Verlag, New York, 1966.

Ia. Eilenberg, S., and Mac Lane, S., On homology groups of infinite complexes and compacta, *in* S. Lefschetz "Algebraic Topology," Appendix A, pp., 344–349. Amer Math. Soc. Providence, Rhode Island, 1942.

Ib. Eilenberg, S., and Mac Lane, S., Infinite cycles and homologies, *Proc. Nat. Acad. Sci. U.S.A.* **27** (1941), 535–539.

I. Eilenberg, S., and Mac Lane, S., Group extensions and homology, *Ann. of Math.* **43** (1942), 758–831.

IIa. Eilenberg, S., and Mac Lane, S., Natural isomorphisms in group theory, *Proc. Nat. Acad. Sci. U.S.A.* **28** (1942), 537–543.

II. Eilenberg, S., and Mac Lane, S., General theory of natural equivalences, *Trans. Amer. Math. Soc.* **58** (1945), 231–294.

IIIa. Eilenberg, S., and Mac Lane, S., Relations between homology and homotopy groups, *Proc. Nat. Acad. Sci. U.S.A.* **29** (1943), 155–158.

III. Eilenberg, S., and Mac Lane, S., Relations between homology and homotopy groups of spaces, *Ann. of Math.* **46** (1945), 480–509.

IV. Eilenberg, S., and Mac Lane, S., Cohomology theory in abstract groups, I, *Ann. of Math.* **48** (1947), 51–78.

V. Eilenberg, S., and Mac Lane S., Cohomology theory in abstract groups, II: Group extensions with a non-abelian kernel, *Ann. of Math.* **48** (1947), 326–341.

VI. Eilenberg, S., and Mac Lane, S., Algebraic cohomology groups and loops, *Duke Math. J.* **14** (1947), 435–463.

VII. Eilenberg, S., and Mac Lane, S., Cohomology and Galois theory, I: Normality of algebras and Teichmüller's cocycle, *Trans. Amer. Math. Soc.* **64** (1948), 1–20.

VIII. Eilenberg, S., and Mac Lane, S., Homology of spaces with operators, II, *Trans. Amer. Math. Soc.* **65** (1949), 49–99.

IX. Eilenberg, S., and Mac Lane, S., Relations between homology and homotopy groups of spaces, II, *Ann. of Math.* **51** (1950), 514–533.

X. Eilenberg S., Mac Lane S., Cohomology theory of abelian groups and homotopy theory, *Proc. Nat. Acad. Sci. U.S.A.* I, **36** (1950), 443–447; II, **36** (1950), 657–663; III, **37** (1951), 307–310; IV, **38** (1952), 325–329. (Opus numbers Xa–Xd.)

Xe. Eilenberg, S., and Mac Lane S., Cohomology theory of abelian groups, by S. Mac Lane, *Proc. Int. Congr. Math., 2, 1950, Cambridge* pp. 8–14.

Xf. Eilenberg, S., and Mac Lane, S., Homotopy groups and algebraic homology theories, by S. Eilenberg, *Proc. Int. Congr. Math. 2, 1950, Cambridge* pp. 350–353.

X. Eilenberg, S., and Mac Lane, S., Homology theories for multiplicative systems, *Trans. Amer. Math. Soc.* **71** (1951), 294–330.

XI. Eilenberg, S., and Mac Lane, S., Acyclic models, *Amer. J. Math.* **75** (1953), 189–199.

XII. Eilenberg, S., and Mac Lane, S., On the groups $H(\Pi, n)$, I, *Ann. of Math.* **58** (1953), 55–106.

XIII. Eilenberg, S., and Mac Lane, S., On the groups $H(\Pi, n)$, II, *Ann. of Math.* **60** (1954), 49–139.

XIV. Eilenberg, S., and Mac Lane, S., On the groups $H(\Pi, n)$, III: Operations and obstructions, *Ann. of Math.* **60** (1954), 513–557.

XV. Eilenberg, S., and Mac Lane, S., On the homology theory of abelian groups, *Canad. J. Math.* **7** (1955), 43–53.

Eilenberg, S., and Miller, E. W., Zero-dimensional families of sets, *Bull. Amer. Math. Soc.* **47** (1941), 921–923.

Eilenberg, S., and Montgomery, D., Fixed point theorems for multi-valued transformations, *Amer. J. Math.* **68** (1946), 214–222.

Eilenberg, S., and Moore, J., Adjoint functors and triples, *Illinois J. Math.* **9** (1965), 381–398.

Eilenberg, S., and Steenrod, N. E., Axiomatic approach to homology theory, *Proc. Nat. Acad. Sci. U.S.A.* **31** (1945), 117–120.

Eilenberg, S., and Steenrod, N. E., "Foundations of Algebraic Topology." Princeton Univ. Press, Princeton, New Jersey, 1952.

Eilenberg, S. and Wilder, R. L., Uniform local connectedness and contractibility, *Amer. J. Math.* **64** (1942), 613–622.

Eilenberg, S. and Zilber, J. A., Semi-simplicial complexes and singular homology, *Ann. of Math.* **51** (1950), 499–513.

Eilenberg, S., and Zilber, J. A., On products of complexes, *Amer. J. Math.* **75** (1953), 200–204.

Freudenthal, H., Der Einfluss der Fundamentalgruppe auf die Bettischen Gruppen, *Ann. of Math.* **47** (1946), 274–316.

Hamsher, R., Eilenberg–Mac Lane algebras and their computation. Ph.D. Thesis, Univ. of Chicago, Chicago, Illinois, 1973.

Hopf, H., Fundamentalgruppe und zweite Bettische Gruppe, *Comment. Math. Helv.* **14** (1941/42), 257–309.

Hopf, H., Uber die Bettischen Gruppen, die zu einer beliebigen Gruppe gehören, *Comment. Math. Helv.* **17** (1944/45), 39–79.

Kan, D. M., Adjoint functors, *Trans. Amer. Math. Soc.* **87** (1958), 294–329.

Lawvere, F. W., Quantifiers and sheaves, *Actes du Congrès Int. des Math.* **1**, (1970), 329, Paris, Gauthier–Villars, 1971.

Mac Lane, S., Aerial gunnery problems, Part II of the Final Report, Applied Mathematics Group, Division of War Research, Columbia University, October, 1945.

Mac Lane, S., Duality for groups, *Bull. Amer. Math. Soc.* **56** (1950), 485–516.

Mac Lane, S., and Schilling, O. F. G., Normal algebraic number fields, *Trans. Amer. Math. Soc.* **50** (1941), 295–384.

Schmidt, F. K., and Mac Lane, S., The generation of inseparable fields, *Proc. Nat. Acad. Sci. U.S.A.* **27** (1941), 583–587.

Steenrod, N., Regular cycles of compact metric spaces, *Ann. of Math.* **41** (1940), 833–851.

Teichmuller, O., Uber die sogenannte nicht Commutative Galoische Theorie und die Relation . . . , *Deutsche. Math.* **5** (1940), 138–149.

Tierney, M., Sheaf theory and the continuum hypothesis, *in* "Toposes, Algebraic Geometry and Logic", *Lecture Notes in Math.* **274**, 13–42. Springer-Verlag, Berlin, 1972.

AMS 18–02, 55–02

DEPARTMENT OF MATHEMATICS
UNIVERSITY OF CHICAGO
CHICAGO, ILLINOIS

What It Means for a Coalgebra To Be Simply Connected

JOHN C. MOORE

Throughout this paper it will be assumed that the ground ring R is a principal ideal domain. Algebra will mean positive differential graded supplemented algebra whose underlying graded R-module is free in each degree, and coalgebra will mean connected differential graded algebra whose underlying graded R-module is free in each degree [4, pp. 116, 117]. Further, it will be assumed that the reader is familiar with the loop and classifying construction, and with their standard properties [4].

Now suppose that C is a coalgebra. If $C_1 = 0$, then one could say that C is simply connected, and indeed one sometimes does. However, this really means that C is simply connected on the chain level, and one would like to know what it means for C to be simply connected without it having to be so on the chain level.

Definition. The coalgebra C is n-connected ($n \in \mathbb{Z}^+$) if $\varepsilon\colon H_*(\Omega C) \to R$ induces an isomorphism in degrees strictly less than n.

Note that in the preceding $\Omega(C)$ denotes the loop algebra of C. Now C is simply connected (1-connected) if $H_0(\Omega C) = R$, i.e., if the fundamental algebra of C is trivial.

Theorem 1. If $n \in \mathbb{Z}$, $n > 0$, and C is an n-connected coalgebra, then there is a morphism of coalgebras $g: C \to D$ such that

(1) $D_q = 0$ for $0 < q \leq n$, and
(2) $\Omega(g): \Omega(C) \to \Omega(D)$ is a chain equivalence.

Note that condition (2) above implies that g is a chain equivalence, but it is considerably stronger than asserting that g is a chain equivalence Indeed suppose that Π is a group such that $H_q(\Pi: \mathbb{Z}) = 0$ for $q \neq 0$ and that C is the normalized chain coalgebra of $K(\Pi, 1)$ with coefficients in R. Now the natural morphism $C \to R$ is a chain equivalence, but $H_0(\Omega C)$ is the group algebra of Π with coefficients in R.

In order to prove the theorem above, a technical result will be proved first. The theorem itself will then follow easily as will some other useful results. In the proofs the notation and results of [4] will be used freely.

Recall that a morphism of coalgebras $f: C' \to C$ is a deformation cofibration if the underlying morphism of graded modules is a split monomorphism such that $\Omega(f)$ induces a chain equivalence and there is a morphism of algebras $g: \Omega(C) \to \Omega(C')$ such that $g\Omega(f) = \Omega(C')$ [4, p. 173].

Lemma. If $i: A' \to A$ is a morphism of algebras such that i induces a chain equivalence that is a split monomorphism of graded modules, then $B(i): B(A') \to B(A)$ is a deformation cofibration.

Proof. Let $j: T(X) \to A$ be a morphism of algebras such that X is a contractible projective module, $T(X)$ the tensor algebra of X, and $i \perp j$: $A' \amalg T(X) \to A$ a fibration of algebras [4, p. 169]. If $g: N \to A' \amalg T(X)$ is the kernel of $i \perp j$, there is a morphism of coalgebras $h: B(A' \amalg T(X)) \to B(N)$ such that $hB(g) = B(N)$ and the composite of h with the natural injection $B(A') \to B(A' \amalg T(X))$ is trivial. Now $B(i \perp j) \top h$ is an isomorphism of $B(A' \amalg T(X))$ with $B(A) \Pi B(N)$ [4, p. 169, 2.5]. Thus there is a morphism of coalgebras $B(A) \to B(A' \amalg T(X))$. Let $u: B(A) \to B(A')$ be the composite of $B(A' \perp *): B(A' \amalg T(X)) \to B(A')$ with this morphism, where $*: T(X) \to A'$ is the trivial morphism, and observe that $uB(i) = B(A')$. It follows that $B(i)$ is a deformation cofibration.

When A is an algebra, $I(A)$ will denote its augmentation ideal considered as a differential graded module.

Technical Theorem. If A is an algebra and $I(A)$ is isomorphic with $X' \oplus X''$ where X'' is contractible, then there is a deformation cofibration of coalgebras $i: C \to B(A)$ such that filtering using coalgebra filtration $E^0(i)$ is a chain equivalence and $E^0(C)$ is isomorphic with the tensor coalgebra of the suspension of X'.

Proof. Having chosen an isomorphism of $I(A)$ with $X' \oplus X''$ and filtered $B(A)$ by its coalgebra filtration, there is a natural isomorphism of $E^0(B(A))$ with the tensor coalgebra $T'(sX' \oplus sX'') = T'(sX') \Pi T'(sX'')$. Since sX'' is contractible, there is a morphism of coalgebras $f: B(A) \to T'(sX'')$, which on the E^0 level becomes the projection of the product coalgebra on its second factor. Let $i: C \to B(A)$ be the kernel of f in the category of coalgebras. Now $E^0(C)$ is isomorphic with $T'(sX')$ and $E^0(i)$ may be viewed as the natural injection in the product.

Filtering $\Omega(C)$ and $\Omega B(A)$ by the filtrations appropriate for derived functors of the second kind, $E^0(\Omega i)$ becomes the natural equivalence of differential modules $R \oplus X' \to A$, and $\Omega(i)$ satisfies the condition of the preceding lemma. The remainder of the proof now follows readily.

Proof of Theorem 1. Since C is n-connected, one has that $I\Omega(C)$ is isomorphic with $X' \oplus X''$ where X'' is contractible and $X_q' = 0$ for $q \leq n - 1$. Let $i': D' \to B\Omega(C)$ be the kernel of $f: B\Omega(C) \to T'(sX'')$ as in the preceding theorem, and let $u: B\Omega C \to B\Omega D'$ be such that the diagram

$$\begin{array}{ccc} D' & \xrightarrow{i'} & B\Omega(C) \\ \downarrow{\scriptstyle \beta(D')} & \swarrow{\scriptstyle u} & \\ B\Omega(D') & & \end{array}$$

is commutative, where $\beta(D')$ is the adjunction morphism of the adjointness between algebras and coalgebras. Let $D = B\Omega(D')$ and $g = u\beta(C)$. Now $g: C \to D$ satisfies the stated condition and the theorem follows.

One knows classically that if X is an n-connected space, then its singular complex is homotopy equivalent with its nth Eilenberg subcomplex. Theorem 1 is a weak algebraic substitute for this fact. It shows that an n-connected coalgebra may be replaced by one that is n-connected on the chain level.

Sometimes it is convenient if possible to replace a coalgebra by one of finite type. This is particularly the case when it seems appropriate to deal with the algebra of cochains of a coalgebra.

Definition. A coalgebra C is *reduced* if the differential module $P(C)$ of primitive chains of C is isomorphic with a coproduct of elementary complexes no one of which is acyclic.

Theorem 2. If C is a coalgebra and either $H_* \Omega C$ is of finite type or R is a field, then there is a reduced coalgebra D, and a morphism of coalgebras $f: C \to B\Omega(D)$ such that $\Omega(f)$ is a chain equivalence.

Proof. The hypotheses imply that $I\Omega(C)$ is isomorphic with $X' \oplus X''$ where X' is a coproduct of elementary complexes and X'' is contractible. Applying the technical theorem, the result follows easily.

Notice that if C is a reduced coalgebra and $H_*(C)$ is of finite type, then C is of finite type. If $H_* \Omega C$ is of finite type, then so also is $H_* C$, and the converse is true if C is simply connected, as was proved by Serre in his thesis. Note that if R is a field, an elementary complex has zero differential, and in this case when a reduced coalgebra C is filtered by its coalgebra filtration $d^0 = 0$, and $E^0(C) = E^1(C)$. If R is a local ring, then the reduction of an elementary complex to a complex over the residue field is an elementary complex, and the reduction of a reduced coalgebra to a coalgebra over the residue field is again a reduced coalgebra.

REFERENCES

[1] J. F. Adams, On the cobar construction, *Proc. Nat. Acad. Sci. U.S.A.* **42** (1956), 409–412.

[2] H. Cartan, Algèbres d'Eilenberg–MacLane et homotopie, Séminaire 1954/55, E.N.S., Paris.

[3] S. Eilenberg and S. Mac Lane, On $H_*(\Pi, n)$ I, II, *Ann. of Math.* **58** (1953), 55–106; **60** (1954), 49–139.

[4] D. Husemoller, J. C. Moore, and J. Stasheff, Differential homological algebra and homogeneous spaces, *J. Pure Appl. Algebra* **5** (1974), 113–185.

This research was partially supported by NSF Grant GP-43903.

AMS 55J99, 18G99

DEPARTMENT OF MATHEMATICS
PRINCETON UNIVERSITY
PRINCETON, NEW JERSEY

Local Complexity of Finite Semigroups

JOHN RHODES and BRET TILSON

1. Complexity Axioms

The complexity function $c\colon \mathbf{TS} \to N$, which assigns to every finite transformation semigroup X a nonnegative integer Xc, is the largest function $d\colon \mathbf{TS} \to N$ satisfying

(A1) (Domination) If $X \prec Y$, then $Xd \leq Yd$
(A2) (Composition) $(Y \circ X)d \leq Yd + Xd$
(A3) (Parallel Composition) $(Y \times X)d \leq \sup\{Yd, Xd\}$
(A4) (Elementary Operations) $X^{\bullet}d = X^{\square}d = \overline{X}d = Xd$
(A5) (Initial Condition) $\mathbf{0}d = 0$
(A6) (Continuity) Let $\mathbf{X} = \{X \in \mathbf{TS} \mid Xd \leq 1\}$. Then $\mathbf{TS} = [\mathbf{X}]$.

Here we adopt the notation and language of the book "Automata, Languages, and Machines," Volume B, by Eilenberg ([2]), where **TS** is the class of all finite transformation semigroups, $\prec$ is division, $\circ$ denotes the wreath product, **0** is the empty ts (ts is the abbreviation of transformation semigroup), and **[X]** indicates the smallest class of ts's containing **X** that is closed under wreath products and division. The elementary operations of (A4) will be explained shortly.

The reader is directed to the book [2] and especially to Chapters XI and XII, written by Tilson, for a complete exposition of the theory of the complexity of transformation semigroups. We will assume that the notation and results therein are known and will make use of them freely. When, for example, Corollary I, 9.7 is referred to herein, we will mean Corollary 9.7 of Chapter I of [2].

Any function d: **TS** $\to N$ that satisfies (A1)–(A6) will be called a *general complexity function*, and (A1) through (A6) themselves will be called *axioms for complexity*. These axioms are motivated by the strong connection between finite state sequential machines and transformation semigroups. For example, see Chapters XI and XII of [1] and Chapter VI of [2].

The meaning of $X \prec Y$ for machines is that Y is capable of computing whatever X can compute. Thus (A1) demands that the "complexity" of Y must be at least as large as the "complexity" of X. The wreath product $Y \circ X$ has as its analog the series composition of machines. Axiom (A2) requires that the series connection of machines be no more "complex" than the sum of the "complexities" of its constituents. (A3) adds the requirement that parallel composition shall not increase the "complexity" beyond that of the constituents. These first three axioms lay the ground rules for what we are trying to measure. A machine cannot be less "complex" than a machine it dominates; under series composition the "complexity" cannot rise faster than the sum of the parts; parallel composition does not increase "complexity" at all.

To discuss (A4) we need to explain the meaning of the operations $X^\bullet$, $X^\square$, and $\overline{X}$. Let $X = (Q, S)$ be a ts. Then Q is a finite set and S is a subsemigroup of $PF(Q)$, the monoid of all partial functions on Q with composition as product. Then $X^\bullet = (Q, S \cup 1_Q)$, where 1_Q is the identity function on Q. If $1_Q \in S$, then $X^\bullet = X$. In terms of machines this operation allows the addition of a neutral input, i.e., an input that does not change the states or the output. According to (A4), this addition should not make the machine more complicated.

$X^\square$ is the ts $(Q \cup \square, S)$, where $\square$ is a new state and the action of S on $Q \cup \square$ is given by

$$q \cdot s = \begin{cases} qs & \text{if } q \in Q \text{ and } qs \neq \varnothing \\ \square & \text{otherwise} \end{cases}$$

The new state $\square$ is called a *sink* state. Notice that the transformations of $X^\square$ are all functions; such a ts is called *complete*. (A4) requires that adding a sink state to a machine not make it more "complex."

The operation $\overline{X}$ is called the *closure* of X; it consists of adding inputs that will reset the machine to any desired state. This can be viewed as giving the machine the ability to be turned off and on. In terms of ts's, $\overline{X} = (Q, S')$ where S' is the subsemigroup of $PF(Q)$ generated by S and

$\{\tilde{q}: q \in Q\}$, $\tilde{q}: Q \to Q$ being the constant function with value q. (A4), then, also requires that this operation does not increase "complexity."

(A5) specifies that the simplest machine, the one with no states and no inputs, be assigned a "complexity" value of zero. (A6) is the requirement that any machine can be built up through composition and division from machines that have "complexity" less than or equal to one. It follows easily from (A1), (A2), and (A6) that if there is a ts of "complexity" n, then there are ts's of each "complexity" less than n.

Because of the Krohn–Rhodes decomposition theorem (discussed below), it will be shown that (A5) and (A6) can be replaced by

(A5′) If X is aperiodic, then $Xd = 0$
(A6′) If X is a transformation group (tg), then $Xd \leq 1$

Recall that a semigroup is aperiodic (or combinatorial) if it has no nontrivial groups as subsemigroups; $X = (Q, S)$ is aperiodic iff S is aperiodic. $X = (Q, S)$ is a tg iff $Q \neq \varnothing$, S is a group, and $1_Q \in S$. If Q is a set, then Q is also treated as the ts $(Q, \varnothing)$. For each integer $n \geq 0$, we denote by $\mathbf{n}$ the set $\{0, \ldots, n-1\}$. Therefore $\mathbf{n}$ is also a ts with n states and no transformations. A particularly important ts is $\bar{\mathbf{2}}^\bullet = (\{0, 1\}, \{\tilde{0}, \tilde{1}, 1\})$.

The Krohn–Rhodes theorem says that each ts $X = (Q, S)$ admits a decomposition

$$X \prec X_1 \circ \cdots \circ X_n$$

where for each index $1 \leq i \leq n$ either $X_i = \bar{\mathbf{2}}^\bullet$ or X_i is a simple group and $X_i \prec S$. As a corollary, if X is aperiodic, then none of the X_i can be simple groups, so

$$X \prec \bar{\mathbf{2}}^\bullet \circ \cdots \circ \bar{\mathbf{2}}^\bullet \tag{1.1}$$

See Chapter II of [2] for an elementary proof of this theorem.

We now prove that (A5) and (A6) can be replaced by (A5′) and (A6′). (A5′) certainly implies (A5). The Krohn–Rhodes theorem implies that

$$\mathbf{TS} = [\mathbf{TG} \cup \bar{\mathbf{2}}^\bullet]$$

where $\mathbf{TG}$ stands for all transformation groups. Therefore (A6) follows from (A5′) and (A6′).

To establish (A5′) we need only show, in view of (1.1), (A1), and (A2), that $\bar{\mathbf{2}}^\bullet d = 0$. But $\mathbf{0}^\square = \mathbf{1}$ and $\mathbf{1}^\square = \mathbf{2}$. Therefore (A4) and (A5) imply $\bar{\mathbf{2}}^\bullet d = 0$.

For (A6′), let $X = (Q, G)$ be a tg. Then it is easy to show that

$$X \prec Q^\bullet \times G$$

(see Corollary I, 9.7).

Since $Q^\bullet$ is obtainable from $\mathbf{0}$ by the elementary operations of (A4), we have $Xd \le Gd$, using (A1), (A3), (A4), and (A5).

Now we claim that any group G is a subgroup of a simple group. Consider the tg $(G \cup G', G)$, where $G \cup G'$ is a disjoint union of two copies of G and the action of G on $G \cup G'$ is

$$x \cdot g = \begin{cases} xg \in G & \text{if } x \in G \\ xg \in G' & \text{if } x \in G' \end{cases}$$

Let card $G = n$. Then each $g \in G$ viewed as a permutation on $2n$ letters is an even permutation. Thus G is a subgroup of the alternating group on $2n$ letters, which is simple unless $n = 2$. But Z_2 is simple, so the assertion is proven.

Since $G \prec P$, a simple group, it suffices to prove $Pd \le 1$. It follows from Theorem V, 11.1 that if

$$P \prec X_1 \circ \cdots \circ X_n \tag{1.2}$$

where $X_i = (Q_i, S_i)$ are complete, then $P \prec S_i$ for some $1 \le i \le n$. By Proposition I, 9.8,

$$S_i \prec X_i \times \cdots \times X_i$$

so using (A1) and (A3) we have $Pd \le X_i d$ for some $1 \le i \le n$. Now using (A6), P admits a decomposition

$$P \prec X_1 \circ \cdots \circ X_n$$

with $X_i d \le 1$. But $X_i \prec X_i^\square$, and $X_i^\square$ is complete, so

$$P \prec X_1^\square \circ \cdots \circ X_n^\square$$

is a decomposition satisfying (1.2). Then $Pd \le X_i^\square d \le 1$, by (A4).

Therefore we have established

Proposition 1.1. Axioms (A1)–(A6) are equivalent with axioms (A1)–(A4), (A5′), and (A6′) ∎

In Chapter XII of [2] it is shown that the complexity function $c\colon \mathbf{TS} \to N$ is the largest function to satisfy (A1)–(A6). The purpose of this paper is to show the existence of another nontrivial function that satisfies (A1)–(A6).

Let $X = (Q, S)$ be a ts and let $e \in S$ be an idempotent. Then the ts $X_e = (Qe, eSe)$ is called a *localization* of X. A long-standing question recently resolved by Rhodes in [3] was whether or not Xc is determined by the complexity of the localizations of X. That is, does

$$Xc = \sup\{X_e c \mid e \text{ an idempotent transformation of } X\}$$

hold? Rhodes answered this in the negative. This question was of interest because of a construction called the derived semigroup of a morphism. A morphism $\varphi\colon S \to T$ in this setting is a *relation* satisfying

$$\text{domain } \varphi = S \tag{1.3}$$

$$(s_1\varphi)(s_2\varphi) \subset (s_1s_2)\varphi \qquad \text{for all } s_1, s_2 \in S \tag{1.4}$$

If, in addition, $\varphi\colon S \to T$ is a function, we call φ a *functional morphism.*

If $\varphi\colon S \to T$ is a morphism, then Chapter XII, Section 8 of [2] defines the derived semigroup Φ of φ, proves

$$S \prec \Phi \circ T \tag{1.5}$$

and shows that the localizations of Φ are determined in a nice way by the nature of φ and special subsemigroups of T. If complexity depended only on localizations, this would lead to a pleasant "kernel" theorem for complexity. These notions will be made more explicit later. Since, however, complexity fails to have this nice property, we asked whether there is a nontrivial general complexity function $d\colon \mathbf{TS} \to N$ that *does* satisfy the localization property, i.e.,

(A7) (Localization) $Xd = \sup\{X_e d \mid e \text{ an idempotent of } X\}$

Any function $d\colon \mathbf{TS} \to N$ satisfying (A1)–(A7) will be called a *local complexity function.* Herein we describe a function $l\colon \mathbf{TS} \to N$ and show it to be the largest local complexity function. A variation of this function was first introduced in [5] and shown to be a lower bound to complexity. That this function satisfies (A2) and (A7) was not recognized at that time.

2. Elementary and Type V Subsemigroups

Following [2], a family of semigroups $\mathbf{V}$ is an $\mathbf{S}$-*variety* iff

$$S \prec T \text{ and } T \in \mathbf{V} \text{ implies } S \in \mathbf{V} \tag{2.1}$$

$$S, T \in \mathbf{V} \text{ implies } S \times T \in \mathbf{V} \tag{2.2}$$

An $\mathbf{S}$-variety of importance here is $\mathbf{R_S}$, the family of all semigroups S characterized by

$$sS = tS \text{ implies } s = t \tag{2.3}$$

Equivalently, $S \in \mathbf{R_S}$ iff each $\mathscr{R}$-class (or transitivity class) of S is a singleton. Equivalently, $S \in \mathbf{R_S}$ iff

$$S \prec U_1 \circ \cdots \circ U_1 \tag{2.4}$$

where $U_1 = 0^I$ is the prime monoid consisting of a zero and a unit. In particular, every member of $\mathbf{R_S}$ is aperiodic. Refer to Chapter V of [2] for more details.

Let S be a semigroup and let

$$\eta: \hat{S} \to S$$

be the Rhodes expansion of S as defined and developed in Chapter XII of [2]. The Rhodes expansion plays a vital role in the major theorems of complexity, and its properties will be used fully in this paper.

For each $s \in \hat{S}$, we consider the stablizer subsemigroup

$$\hat{S}_s = \{w \in \hat{S} \mid sw = s\}$$

A subsemigroup W of S is called an *elementary subsemigroup* of S if

$$W \subset \hat{S}_s \eta$$

for some $s \in \hat{S}$. The set of elementary subsemigroups of S will be denoted $S_{\mathbf{E}}$. If $S \in S_{\mathbf{E}}$, i.e., if S is an elementary subsemigroup of itself, then S is called an *elementary semigroup*.

The only elementary subsemigroup of a group is the identity; this follows from the fact that $\hat{G} \approx G$ if G is a group. The semigroup $U_1 = 0^I$ is elementary as is any set of idempotents $\{e_1, \ldots, e_n\}$ with $e_i e_j e_i = e_j$ if $j \geq i$. These statements follow easily from the construction of the Rhodes expansion.

It is shown in Section XII, 12 of [2] that $\hat{S}_s \in \mathbf{R_S}$ for each $s \in \hat{S}$. Since η is a functional morphism, we have

$$\text{If } W \in S_{\mathbf{E}}, \text{ then } W \in \mathbf{R_S} \tag{2.5}$$

This by no means fully characterizes elementary subsemigroups of S; for example, the elements of $W \in S_{\mathrm{E}}$ must form a chain in the $\mathscr{L}$ ordering of S.

If $\varphi: S \to T$ is a functional morphism, then there is a functional morphism $\hat{\varphi}: \hat{S} \to \hat{T}$ satisfying $\hat{\varphi}\eta = \eta\varphi$, i.e., the diagram

$$\begin{array}{ccc} \hat{S} & \xrightarrow{\hat{\varphi}} & \hat{T} \\ \downarrow{\scriptstyle \eta} & & \downarrow{\scriptstyle \eta} \\ S & \xrightarrow{\varphi} & T \end{array}$$

commutes. It follows easily that

$$\text{If } W \in S_{\mathrm{E}} \text{ and } \varphi: S \to T \text{ is a functional morphism, then } W\varphi \in T_{\mathrm{E}} \tag{2.6}$$

$$\text{If } W \in S_{\mathrm{E}} \text{ and } S \subset T, \text{ then } W \in T_{\mathrm{E}} \tag{2.7}$$

Let $\mathbf{V}$ be an $\mathbf{S}$-variety. A subsemigroup S' of S is a *type* $\mathbf{V}$ *subsemigroup of* S if for each morphism

$$\varphi : S \to T$$

with $T \in \mathbf{V}$, there exists an elementary subsemigroup W of T such that

$$S' \subset W\varphi^{-1}$$

Recall that a morphism is a relation satisfying (1.3) and (1.4).

Let $S_{\mathbf{V}}$ denote the set of all type $\mathbf{V}$ subsemigroups of S. If $S \in S_{\mathbf{V}}$, we call S a *type* $\mathbf{V}$ *semigroup.*

$T \subset S$ is a *maximal* type $\mathbf{V}$ subsemigroup of S if $T \in S_{\mathbf{V}}$ and $T \subset W \subset S$ with $W \in S_{\mathbf{V}}$ implies $T = W$.

The following facts are easily verified.

$$W \in S_{\mathbf{V}} \text{ and } S \subset T \text{ implies } W \in T_{\mathbf{V}} \tag{2.8}$$

$$W \in S_{\mathbf{V}} \text{ and } T \subset W \text{ implies } T \in S_{\mathbf{V}} \tag{2.9}$$

Theorem 2.1. Let $\mathbf{V}$ be an $\mathbf{S}$-variety, and let S be a semigroup. Then there exists a morphism

$$\varphi : S \to T, \qquad T \in \mathbf{V}$$

such that

$$W \in T_{\mathbf{E}} \text{ implies } W\varphi^{-1} \in S_{\mathbf{V}}$$

Furthermore, all maximal type $\mathbf{V}$ subsemigroups of S are of this form.

Proof. Consider the set of all morphisms φ: $S \to T$, with $T \in \mathbf{V}$. Call two such morphisms φ: $S \to T$ and φ': $S \to T'$ equivalent if

$$\{W\varphi^{-1} : W \in T_{\mathbf{E}}\} = \{W'\varphi'^{-1} : W' \in T'_{\mathbf{E}}\} \tag{2.10}$$

Equation (2.10) clearly defines an equivalence relation on this set of morphisms that has only a finite number of equivalence classes. Let

$$\varphi_i : S \to T_i, \qquad i = 1, \ldots, n$$

be a complete set of representatives of the classes. Define φ: $S \to T_1 \times \cdots \times T_n$ by

$$s\varphi = \{(t_1, \ldots, t_n) : t_i \in s\varphi_i,\ i = 1, \ldots, n\}$$

It is easy to verify that φ is a morphism. We will show that φ is the desired morphism. First, note that $T_1 \times \cdots \times T_n \in \mathbf{V}$. Secondly, if $(t_1, \ldots, t_n) \in T_1 \times \cdots \times T_n$, then

$$\begin{aligned}(t_1, \ldots, t_n)\varphi^{-1} &= \{s \in S : s \in t_i\varphi_i^{-1},\ i = 1, \ldots, n\} \\ &= \bigcap_{i=1}^{n} t_i\varphi_i^{-1}\end{aligned}$$

Let W be an elementary subsemigroup of $T_1 \times \cdots \times T_n$. If $\pi_i: T_1 \times \cdots \times T_n \to T_i$ is the projection, then by (2.6), $W\pi_i$ is an elementary subsemigroup of T_i. Now if $s \in W\varphi^{-1}$, then there exists $(t_1, \ldots, t_n) \in W$ such that $s \in (t_1, \ldots, t_n)\varphi^{-1}$. Therefore

$$s \in \bigcap_{i=1}^{n} t_i \varphi_i^{-1}$$

But $t_i \in W\pi_i$, so $s \in \bigcap_{i=1}^{n} W\pi_i\varphi_i^{-1}$, and we have

$$W\varphi^{-1} \subset \bigcap_{i=1}^{n} W\pi_i\varphi_i^{-1}$$

We need to show that $W\varphi^{-1}$ is a type $\mathbf{V}$ subsemigroup of S. Let

$$\psi: S \to T', \qquad T' \in \mathbf{V}$$

be any morphism. Then for some $j = 1, \ldots, n$, we have $\psi: S \to T'$ equivalent with $\varphi_j: S \to T_j$. Therefore there exists $W' \in T'_{\mathbf{E}}$ such that

$$W'\psi^{-1} = W\pi_j\varphi_j^{-1}$$

Hence

$$W\varphi^{-1} \subset \bigcap_{i=1}^{n} W\pi_i\varphi_i^{-1} \subset W\pi_j\varphi_j^{-1} = W'\psi^{-1}$$

Therefore $W\varphi^{-1}$ is a type $\mathbf{V}$ subsemigroup of S.

Finally, if S' is a maximal type $\mathbf{V}$ subsemigroup of S, then there exists an elementary subsemigroup W of $T_1 \times \cdots \times T_n$ such that

$$S' \subset W\varphi^{-1}$$

But $W\varphi^{-1}$ is a type $\mathbf{V}$ subsemigroup of S, so by the maximality of S', we have

$$S' = W\varphi^{-1} \quad \blacksquare$$

For convenience, we will call any morphism $\varphi: S \to T$ that satisfies the conditions of Theorem 2.1 a *type* $\mathbf{V}$ *morphism of* S.

Proposition 2.2. If $S' \in S_{\mathbf{V}}$ and $\varphi: S \to T$ is a morphism, then there exists $T' \in T_{\mathbf{V}}$ such that

$$S' \subset T'\varphi^{-1}$$

Furthermore, if S' is a type $\mathbf{V}$ semigroup, then T' may be chosen to be a type $\mathbf{V}$ semigroup.

Proof. Let $\psi: T \to V$ be a type $\mathbf{V}$ morphism of T. Then

$$\varphi\psi: S \to V, \qquad V \in \mathbf{V}$$

so there exists $W \in V_E$ such that

$$S' \subset W(\varphi\psi)^{-1} = W\psi^{-1}\varphi$$

By Theorem 2.1, we have $W\psi^{-1} \in T_V$. Let $T' = W\psi^{-1}$.

Now let S' be a type **V** semigroup, i.e., $S' \in S'_V$. Choose $T' \in T_V$ as small as possible subject to the condition

$$S' \subset T'\varphi^{-1}$$

Define the restriction morphism $\varphi' : S' \to T'$ by

$$s\varphi' = s\varphi \cap T'$$

Since S' is a type **V** semigroup, there must exist $T'' \in T'_V$ such that $S' \subset T''\varphi'^{-1}$. But

$$T''\varphi'^{-1} \subset T''\varphi^{-1}$$

so by the minimality of T', T'' must equal T'. This proves that $T' \in T'_V$, i.e., that T' is a type **V** semigroup ▮

Corollary 2.3. If $\varphi: S \to T$ is a functional morphism and $S' \in S_V$, then $S'\varphi \in T_V$. Furthermore, if S' is a type **V** semigroup, then so is $S'\varphi$.

Proof. If $S' \in S_V$, there exists $T' \in T_V$ such that

$$S' \subset T'\varphi^{-1}$$

But since φ is a function, we have

$$S'\varphi \subset T'\varphi^{-1}\varphi \subset T'$$

Therefore $S'\varphi \in T_V$ follows from (2.9).

If S' is a type **V** semigroup, consider the restriction of φ

$$\varphi' : S' \to S'\varphi$$

Then $S'\varphi' = S'\varphi \in (S'\varphi)_V$, i.e., $S'\varphi$ is a type **V** semigroup ▮

We now particularize the **S**-variety **V** to the two basic classes of semigroups singled out by the Krohn–Rhodes decomposition theorem, namely, aperiodic semigroups and groups.

If **V** is the **S**-variety of all aperiodic semigroups, then type **V** will be called type I (read as type one) and S_V will be denoted by S_I.

If **V** is the **S**-variety of all groups, then type **V** will be called type II (read as type two) and S_V will be denoted by S_{II}. Since the only elementary subsemigroup of a group is its identity, and Theorem 2.1 guarantees the existence of a type II morphism

$$\varphi: S \to G$$

it follows that S has a unique maximal type II subsemigroup, namely, $1\varphi^{-1}$. We shall, by an abuse of notation, call this subsemigroup S_{II}.

Note that there need not be a unique maximal type I subsemigroup of S.

Type I and type II subsemigroups are also defined and used in [5], and their definitions differ from those of this chapter. However, we now show they are equivalent. The definition for type I involves R_1 semigroups. A semigroup S is R_1 iff $eS = fS$, and e and f are idempotents implies $e = f$.

In [5] the definitions are:

S' is a type I subsemigroup of S iff

For each morphism $\varphi: S \to A$ with A aperiodic there exists an R_1 semigroup R contained in A such that $S' \subset R\varphi^{-1}$. (2.11)

S' is a type II subsemigroup of S iff

For each morphism $\varphi: S \to G$ with G a group $S' \subset 1\varphi^{-1}$ holds. (2.12)

Proposition 2.4. (a) S' is a type I subsemigroup of S iff (2.11) holds.
(b) S' is a type II subsemigroup of S iff (2.12) holds.

Proof. (a) Since elementary subsemigroups belong to $\mathbf{R}_S$, they are clearly R_1, so S' being type I in S implies (2.11). Assume (2.11) holds and let $\varphi: S \to A$ be a morphism with A aperiodic. We must show the existence of an elementary subsemigroup W of A such that $S' \subset W\varphi^{-1}$. Let $\eta: \hat{A} \to A$ be the Rhodes expansion of A. Then $\hat{A}$ is aperiodic and η is a surjective functional morphism. Considering the morphism $\varphi\eta^{-1}: S \to \hat{A}$, (2.11) implies that there is an R_1 semigroup R in $\hat{A}$ such that $S' \subset R\eta\varphi^{-1}$. Let R' be a minimal right ideal of R. Since R is aperiodic, every element of R' is an idempotent. Since $wR = R'$ for all $w \in R'$ and since R is R_1, it follows that R' is a singleton, i.e., $R' = w$. Therefore $wR = w$ and $R \subset \hat{A}_w$, the stablizer subsemigroup. Hence $R\eta$ is an elementary subsemigroup of A. Letting $W = R\eta$ completes the proof.

(b) Since the only elementary subsemigroup of a group is the identity, (b) follows directly ∎

In view of Proposition 2.4, we are free to quote certain results from [5].

Every group is a type I semigroup. (2.13)

A semigroup generated by its idempotents is type II. (2.14)

If S is a nonaperiodic type I semigroup and T is a type II subsemigroup of S, then $Tc < Sc$. (2.15)

The reader is referred to [4] and [5] for further information about these types of subsemigroups.

Further facts of interest are

The only type II subsemigroup of a group is the identity. (2.16)

If T is a type I subsemigroup of an aperiodic semigroup S, then $T \in S_{\mathbf{E}}$. (2.17)

Both these facts follow from

Proposition 2.5. If $S' \in S_{\mathbf{V}}$ and $S \in \mathbf{V}$, then $S' \in S_{\mathbf{E}}$.

Proof. Consider the identity morphism $I: S \to S$. There must be a $W \in S_{\mathbf{E}}$ such that $S' \subset WI^{-1} = W$ ∎

Type I and type II subsemigroups will be used in the next section to define the largest local complexity function. It is not known how to construct all type I and type II subsemigroups of a semigroup S, but [5] is devoted to the construction of a wide class of these subsemigroups. In fact, if S is regular, S_{II} is completely constructible using [5]. Furthermore, it can be determined whether or not S_{II} is aperiodic, which is of interest because of Theorem 3.1 of [5]:

$$\text{If } S_{\mathrm{II}} \text{ is aperiodic, then } Sc \leq 1$$

3. The Largest Local Complexity Function

Let S be a semigroup and let

$$S = U_0 \supset T_1 \supset U_1 \supset \cdots \supset T_n \supset U_n \tag{3.1}$$

be a chain of subsemigroups of S satisfying

Each T_i, $i = 1, \ldots, n$, is a nonaperiodic type I semigroup;

Each U_i, $i = 1, \ldots, n$, is a type II subsemigroup of $(T_i 2)$ (3.2)

Any chain (3.1) will be called an *alternating series of* S of length n.

Using (2.15), it easily follows that if S has a series (3.1), then $n \leq Sc$. Therefore we can define

$$Sl = \sup\{\text{length of alternating series of } S\}$$

and then extend this definition to l: $\mathbf{TS} \to N$ by

$$Xl = Sl \qquad \text{if} \quad X = (Q, S)$$

From (2.15) it follows that l is a lower bound to complexity, that is,

$$Xl \leq Xc$$

It immediately follows that l satisfies (A5′) and (A6′), as can also be seen directly from the definition of l. Also, since the semigroups of X and $X^{\square}$ are isomorphic, we have immediately part of (A4)

$$Xl = X^{\square}l \tag{3.3}$$

Let F_n denote the monoid of all functions on n letters, and let S_n denote the symmetric group on n letters. In [4], it is shown that F_n is a type I semigroup and $F_n - S_n$ is a type II semigroup. Since $F_{n-1} \subset F_n - S_n$, $n \geq 2$, we have an alternating series

$$F_n \supset F_n - S_n \supset F_{n-1} \supset \cdots \supset F_2 \supset F_2 - S_2$$

of length $n - 1$ for F_n, $n \geq 1$. Thus

$$n - 1 \leq F_n l \leq F_n c = n - 1$$

so $F_n l = n - 1 = F_n c$, $n \geq 1$. Therefore l: **TS** $\rightarrow N$ is surjective and agrees with complexity on F_n, $n \geq 1$. However, there are examples of semigroups S in [3] where $Sl < Sc$.

If S is a subsemigroup of T and (3.1) is a series for S, then (3.1) is also a series for T. Therefore

$$\text{If } S \text{ is a subsemigroup of } T, \text{ then } Sl \leq Tl \tag{3.4}$$

Let $\varphi: S \rightarrow T$ be a morphism. Then define the *kernel of* φ to be the set

$$\ker \varphi = \{W\varphi^{-1} \mid W \in T_{\mathbf{E}}\}$$

Thus the kernel of φ is the set of subsemigroups obtained as inverse images of elementary subsemigroups of T under φ. We extend the function l to morphisms by

$$\varphi l = \sup\{S'l \mid S' \in \ker \varphi\}$$

Proposition 3.1. Let $\varphi: S \rightarrow T$ be a morphism. Then

$$Sl \leq \varphi l + Tl$$

Proof. Let $Sl = n > 0$ and let (3.1) be a series for S of length n. By Proposition 2.2, there exists a type I semigroup $T_1' \subset T$ such that

$$T_1 \subset T_1'\varphi^{-1}$$

Let $\varphi_1: T_1 \rightarrow T_1'$ be defined by restriction. Again, by Proposition 2.2, there exists a type II subsemigroup U_1' of T_1' such that

$$U_1 \subset U_1'\varphi_1^{-1}$$

Let $\varphi_1': U_1 \rightarrow U_1'$ be defined by restriction. Continue in this manner choosing the subsemigroups T_i', U_i'.

Let $k \geq 0$ be the smallest integer such that T'_{k+1} is aperiodic. Then

$$T \supset T_1' \supset U_1' \supset \cdots \supset T_k' \supset U_k'$$

is an alternating series for T, so $k \leq Tl$.

Since T'_{k+1} is aperiodic and type I, we have, by (2.17),

$$T'_{k+1} \in T_{\mathbf{E}}$$

Therefore

$$T_{k+1} = T'_{k+1}\varphi_{k+1}^{-1} \subset T'_{k+1}\varphi^{-1} \in \ker \varphi$$

and using (3.4) we have

$$T_{k+1}l \leq T'_{k+1}\varphi^{-1}l \leq \varphi l$$

But

$$T_{k+1} \supset U_{k+1} \supset \cdots \supset T_n \supset U_n$$

is an alternating series for T_{k+1}, so $n - k \leq T_{k+1}l$. It follows that

$$Sl = n = (n - k) + k \leq \varphi l + Tl \quad ∎$$

A morphism $\varphi: S \to T$ is *aperiodic* if $T'\varphi^{-1}$ is aperiodic whenever T' is an aperiodic subsemigroup of T. One of the central theorems of complexity theory states that if $\varphi: S \to T$ is aperiodic, then $Sc \leq Tc$. We have the same result for local complexity.

Corollary 3.2. Let $\varphi: S \to T$ be an aperiodic morphism. Then

$$Sl \leq Tl$$

Proof. Since all elementary subsemigroups of T are aperiodic, we have $\varphi l = 0$ ∎

Corollary 3.3. If X and Y are ts's and $X \prec Y$, then

$$Xl \leq Yl$$

Proof. $(Q, S) \prec (P, T)$ implies the existence of an aperiodic morphism $\varphi: S \to T$. See Exercise XII, 4.6 of [2] ∎

Therefore l satisfies (A1). We next show that l satisfies (A3); it suffices to prove

Proposition 3.4. $(S_1 \times S_2)l \leq \sup\{S_1 l, S_2 l\}$

Proof. We may assume $(S_1 \times S_2)l = n > 0$. Let

$$S_1 \times S_2 \supset T_1 \supset U_1 \supset \cdots \supset T_n \supset U_n$$

be a series for $S_1 \times S_2$ of length n. Let

$$\pi_i: S_1 \times S_2 \to S_i, \quad i = 1, 2$$

be the projections. Then by Corollary 2.3,

$$S_i \supset T_1\pi_i \supset \cdots \supset T_n\pi_i \supset U_n\pi_i$$

are alternating series for S_i, $i = 1, 2$, except for the possibility that the type I terms are aperiodic. But

$$T_n \subset T_n\pi_1 \times T_n\pi_2$$

and T_n is not aperiodic, so one of $T_n\pi_1$ and $T_n\pi_2$ is not aperiodic. Therefore either $S_1 l \geq n$ or $S_2 l \geq n$. The assertion follows ▮

The next two propositions show that l: **TS** $\to N$ satisfies (A4).

Proposition 3.5. Let X be a ts. Then

$$\overline{X}l = Xl$$

Proof. Let $X = (Q, S)$. Then $\overline{X} = (Q, S')$ where S' is the subsemigroup of $PF(Q)$ generated by S and $\{\tilde{q}: q \in Q\}$. Let

$$I = \{t \in S' \,|\, t = t_1\tilde{q}t_2 \text{ for some } q \in Q,\ t_1, t_2 \in S'\}$$

I is an ideal of S'. Let $t = t_1\tilde{q}t_2 \in I$. Then

$$Qt = Qt_1\tilde{q}t_2 \subset Q\tilde{q}t_2 = qt_2$$

Therefore card $Qt \leq 1$. It follows that I is aperiodic. Define $\varphi: S' \to S$ by

$$t\varphi = \begin{cases} t & \text{if } t \in S' - I \\ S & \text{if } t \in I \end{cases}$$

A straightforward calculation shows that φ is an aperiodic morphism. Therefore $\overline{X}l \leq Xl$. The reverse inequality follows from $X \prec \overline{X}$ ▮

Let $X = (Q, S)$. Then the semigroup of $X^\bullet$ is either S or S^I, where $S^I = S \cup \{I\}$, I being a new element satisfying $sI = s = Is$ for all $s \in S$. Therefore, to establish $X^\bullet l = Xl$, it suffices to prove

Proposition 3.6. Let S be a semigroup. Then

$$S^I l = Sl$$

Proof. Since $S \subset S^I$ we need only show that $S^I l \leq Sl$.

Suppose $T \subset S^I$ is a type I semigroup. We show that $T - I$ is a type I semigroup. If $T - I = T$ we are done, so assume $I \in T$. Let

$$\varphi: T - I \to A, \qquad A \text{ aperiodic}$$

be a morphism. Extend φ to $\varphi: T \to A^I$ by setting $I\varphi = I$. Since A^I is aperiodic, there exists a $W \in (A^I)_{\mathbf{E}}$ such that

$$T - I \subset W\varphi^{-1}$$

But I is not needed in W in the above, so

$$T - I \subset (W - I)\varphi^{-1}$$

But W is an R_1 semigroup, so $W - I$ is also R_1. It follows from Proposition 2.4 that $T - I$ is a type I semigroup.

Suppose U is a type II subsemigroup of S^I. Let

$$\varphi: S \to G, \qquad G \text{ a group}$$

be a morphism. Extend φ to $\varphi^I: S^I \to G$ by setting $I\varphi = 1$, the identity of G. Then $U \subset 1(\varphi^I)^{-1}$ and $U - I \subset 1\varphi^{-1}$. Therefore $U - I$ is a type II subsemigroup of S.

It now follows that any alternating series

$$S^I \supset T_1 \supset U_1 \supset \cdots \supset T_n \supset U_n$$

yields an alternating series

$$S \supset T_1 - I \supset U_1 - I \supset \cdots \supset T_n - I \supset U_n - I$$

for S of the same length. Thus $S^I l \leq Sl$ ∎

We now will show that (A2) is satisfied by l: $\mathbf{TS} \to N$. Since the semigroup of a wreath product is expressible as a semidirect product, we first prove a semidirect product version of (A2). The reader is referred to V, 4.

Proposition 3.7. Let $S_2 * S_1$ be a semidirect product of semigroups S_1 and S_2. Then

$$(S_2 * S_1)l \leq S_2 l + S_1 l$$

Proof. Consider the projection morphism

$$\pi: S_2 * S_1 \to S_1$$

Proposition 3.1 yields

$$(S_2 * S_1)l \leq \pi l + S_1 l$$

so it suffices to show that $\pi l \leq S_2 l$.

Every member of ker π is in the form $S_2 * W$, where W is an elementary subsemigroup of S_1, and the left action of W on S_2 is the restriction of the action of S_1 on S_2. This action can be extended to a unitary action of W^I

on ${S_2}^I$. Since ${S_2}^I\, l = S_2\, l$, $W^I \in \mathbf{R}_s$, and $S_2 * W \subset {S_2}^I * W^I$, it will suffice to prove

$$(S_2 * W)l \leq S_2\, l \tag{3.5}$$

under the assumptions that S_2 and W are monoids, the action of W on S_2 is unitary, and $W \in \mathbf{R}_s$. Therefore, by Proposition V, 4.4 and (2.4) we have

$$S_2 * W \prec S_2 \circ W \prec S_2 \circ U_1 \circ \cdots \circ U_1$$

In view of this inequality it suffices, to establish (3.5), to prove

$$(X \circ U_1)l \leq Xl$$

for any complete ts X. If T is the semigroup of X, then the semigroup of $X \circ U_1$ is $(T \times T) * U_1$. Thus, using Proposition 3.4, we see it suffices to establish

$$(S * U_1)l \leq Sl$$

for any semigroup S and any left action of U_1 on S. This we now do. $S * U_1$ is the union of a subsemigroup

$$S_1 = \{(s, 1) | s \in S\} \text{ and an ideal } I = \{(s, 0) | s \in S\}$$

Define two morphisms φ_i: $S * U_1 \to S$, $i = 1, 2$, by

$$(s, x)\varphi_1 = \begin{cases} 1s & \text{if } x = 1 \\ S & \text{if } x = 0 \end{cases}$$

$$(s, x)\varphi_2 = 0s$$

where $1s$ and $0s$ are the results of the left action of 1 and 0 on s. As was shown in Section XII, 7, φ_1 is injective on groups in S_1 and φ_2 is injective on groups in I. It follows that

$$\varphi: S * U_1 \to S \times S$$

$$(s, x)\varphi = (s, x)\varphi_1 \times (s, x)\varphi_2$$

is aperiodic. Therefore $(S * U_1)l \leq Sl$ follows from Corollary 3.2 and Proposition 3.4 ∎

Corollary 3.8. (A2) is satisfied by l: $\mathbf{TS} \to N$. That is

$$(X \circ Y)l \leq Xl + Yl$$

Proof. Since $X \circ Y \prec X^\square \circ Y^\square$ and $Xl = X^\square l$, it suffices to assume that $X = (Q, S)$ and $Y = (P, T)$ are complete. In this case, the semigroup of

$X \circ Y$ is $S^{(n)} * T$, where $n = \text{card } P$ and $S^{(n)}$ is a direct product of n copies of S. Therefore

$$(X \circ Y)l \leq S^{(n)}l + Tl \leq Sl + Tl = Xl + Yl \quad ▮$$

Therefore we have established that l: $\mathbf{TS} \to N$ is a general complexity function. That l is also a local complexity function follows from the next proposition.

Proposition 3.9. Let X be a ts. Then

$$Xl = \sup\{X_e l \mid e \text{ an idempotent of } X\}$$

Proof. Let $X = (Q, S)$. Since the semigroup of X_e is eSe, it suffices to prove

$$Sl = \sup\{(eSe)l \mid e \in S, e^2 = e\}$$

Let E be the set of idempotents in S, and let $T = ESE$. The reduction theorem (Theorem XI, 4.1) implies that

$$S \prec T^{\bullet} \circ A$$

for some aperiodic ts A. Using (A1), (A2), and (A4), we have $Sl \leq Tl$.

Let $\tilde{E} = \{\tilde{e} \mid e \in E\}$ be the semigroup with multiplication $\tilde{e}\tilde{f} = \tilde{f}$ for all $e, f \in E$. Every element of $\tilde{E}$ is idempotent, so $\tilde{E}$ is aperiodic. Define

$$\varphi : T \to \tilde{E}$$

by $t\varphi = \{\tilde{e} \mid t \in Te\}$. φ is clearly a morphism, so by Proposition 3.1, $Tl \leq \varphi l + \tilde{E}l = \varphi l$.

Since the elements of $\tilde{E}$ are mutally $\mathscr{R}$ equivalent, the elementary subsemigroups of $\tilde{E}$ are the singletons. Since $\tilde{e}\varphi^{-1} = Te$, it follows that

$$\ker \varphi = \{Te \mid e \in E\}$$

For each $e \in E$, define ψ_e: $Te \to eTe$ by $t\psi_e = et$. ψ_e is easily seen to be an aperiodic morphism, so

$$(Te)l \leq (eTe)l \leq (eSe)l$$

Then

$$Sl \leq \varphi l = \sup\{(Te)l \mid e \in E\} \leq \sup\{(eSe)l \mid e \in E\}$$

Since $eSe \subset S$, the reverse inequality follows ▮

We now utilize certain results from Chapter XII of [2]. Recall that S_{I} is the set of type I subsemigroups of S and S_{II} is the unique maximal type II subsemigroup of S.

Lemma 3.10. Let d: $\mathbf{TS} \to N$ be a function that satisfies (A1)–(A7). Let $X = (Q, S)$ be a ts. Then

$$Xd \leq \sup\{Td \mid T \in S_{\mathrm{I}}\} \tag{3.6}$$

$$Xd \leq S_{\mathrm{II}}\,d + 1 \tag{3.7}$$

Proof. By Corollary I, 9.4,

$$X \prec S \times A$$

for some aperiodic ts A. Thus, using (A1), (A3), and (A5′), we obtain $Xd \leq Sd$. Therefore, it suffices to prove (3.6) and (3.7) with X replaced by S.

Let $\mathbf{V}$ be an $\mathbf{S}$-variety and let

$$\varphi: S \to V$$

be a type $\mathbf{V}$ morphism of S. Let η: $\hat{V} \to V$ be the Rhodes expansion of V. There results the composition morphism $\varphi\eta^{-1}: S \to \hat{V}$. Proposition XII, 8.1 gives

$$S \prec \Phi \circ \hat{V} \tag{3.8}$$

where Φ is the derived semigroup of $\varphi\eta^{-1}$ as defined in Section XII, 8. In this same section it is shown that for each $v \in \hat{V}$, there is a subsemigroup Φ_v of Φ that is isomorphic with $\hat{V}_v\eta\varphi^{-1}$, the image of the stablizer of v under the inverse of $\varphi\eta^{-1}$. Since $\hat{V}_v\eta \in V_{\mathbf{E}}$ and since $\varphi: S \to V$ is a type $\mathbf{V}$ morphism, each $\hat{V}_v\eta\varphi^{-1}$ is a type $\mathbf{V}$ subsemigroup of S.

It follows easily from the definition of Φ that for every nonzero idempotent $e \in \Phi$, there is a $v \in \hat{V}$ such that

$$e\Phi e \subset \Phi_v \cup \{0\}$$

Since d satisfies (A1) and (A7), we can conclude that

$$\Phi d \leq \sup\{T^0 d \mid T \in S_{\mathbf{V}}\} \tag{3.9}$$

where T^0 denotes the result of adjoining a new zero to T. It is an easy exercise to show that $T \prec T^0 \prec T \times U_1$. Consequently, $T^0 d = Td$. Therefore, combining (3.8) and (3.9) we obtain

$$Sd \leq \sup\{Td \mid T \in S_{\mathbf{V}}\} + \hat{V}d \tag{3.10}$$

Now applying (3.10) to type I and type II we obtain the assertions, since $\hat{A}$ is aperiodic if A is aperiodic, and $\hat{G} \approx G$ ∎

Theorem 3.11. l: **TS** $\to N$ is the largest function that satisfies (A1)–(A7).

Proof. We first establish

$$\text{If } S \text{ is a nonaperiodic type I semigroup then } Sl \geq S_{\text{II}}\, l + 1 \tag{3.11}$$

Let $S_{\text{II}} l = n$ and let

$$S_{\text{II}} = U_0 \supset T_1 \supset \cdots \supset T_n \supset U_n$$

be a series of length n for S_{II}. Since S is a nonaperiodic type I semigroup and S_{II} is a type II subsemigroup of S, the series

$$S \supset S_{\text{II}} \supset T_1 \supset \cdots \supset T_n \supset U_n$$

is an alternating series for S of length $n + 1$. Therefore

$$Sl \geq n + 1 = S_{\text{II}} l + 1$$

We are now ready to prove the theorem. Let d: **TS** $\to N$ be a function satisfying (A1)–(A7), and let $X = (Q, S)$. We induct on card S. If S is not type I, then each type I subsemigroup of S is proper. Then by (3.6) and induction

$$\begin{aligned} Xd &\leq \sup\{Td \mid T \in S_{\text{I}}\} \\ &\leq \sup\{Tl \mid T \in S_{\text{I}}\} \leq Sl = Xl \end{aligned}$$

Therefore we may assume S is type I. If S is aperiodic, then $Xd = 0 = Xl$. If not, then by (2.15) S_{II} is a proper subsemigroup of S. Then, using (3.7), (3.11), and induction,

$$Xd \leq S_{\text{II}}\, d + 1 \leq S_{\text{II}}\, l + 1 \leq Sl = Xl \quad \blacksquare$$

Thus l: **TS** $\to N$ is the largest local complexity function. Since, as shown in [3], c does not satisfy (A7), we see that $l \neq c$. In fact, an unpublished example shows that l and c need not agree even when restricted to monoids [where (A7) becomes vacuous].

REFERENCES

[1] S. Eilenberg, "Automata, Languages, and Machines," Volume A. Academic Press, New York, 1974.

[2] S. Eilenberg, "Automata, Languages, and Machines," Volume B. Academic Press, New York, 1976.

[3] J. Rhodes, Kernel systems—a global study of homomorphisms on finite semigroups, (1974), Preprint.

[4] J. Rhodes and B. Tilson, Lower bounds for complexity of finite semigroups, *J. Pure Appl. Algebra* **1** (1971), 79–95.
[5] J. Rhodes and B. Tilson, Improved lower bounds for the complexity of finite semigroups, *J. Pure Appl. Algebra* **2** (1972), 13–71.

AMS 20M20

John Rhodes
DEPARTMENT OF MATHEMATICS
UNIVERSITY OF CALIFORNIA
BERKELEY, CALIFORNIA

Bret Tilson
DEPARTMENT OF MATHEMATICS
CITY UNIVERSITY OF NEW YORK
QUEENS COLLEGE
FLUSHING, NEW YORK

The Global Dimensions of Ore Extensions and Weyl Algebras

GEORGE S. RINEHART and ALEX ROSENBERG

Introduction

This chapter originated from the study of a theorem of Hart's [9, Th. 2.4], which asserts that if E is a division ring of characteristic zero and $A_1(E) = E[x, t]$ with $tx - xt = 1$ then gl.dim $A_1(E)$ is 1 or 2 with gl.dim $A_1(E) = 2$ if and only if there exists a finite dimensional vector space over E that is a left $A_1(E)$-module.

In Section 1 we consider a left and right noetherian ring R with a derivation and let S be the usual ring of differential polynomials over R. We show that for a left S-module M that is finitely generated as an R-module, $\text{l.dim}_S M = \text{l.dim}_R M + 1$. Combining this with Theorem 3.8 of [14] shows that if gl.dim $R = d < \infty$, then gl.dim $S = d + 1$ if and only if there exists a left S-module M, finitely generated as an R-module with $\text{l.dim}_R M = d$.

In Section 2 we study the Weyl algebras $A_n(R) = R[x_1, \ldots, x_n, t_1, \ldots, t_n]$, $t_i x_i - x_i t_i = 1$, and, for $i \neq j$, $x_i t_j = t_j x_i$. For a two-sided noetherian ring R we show that gl.dim $A_1(R) =$ gl.dim $R + 1$ or 2, with the second case occurring if and only if there exists a left $A_1(R)$-module M that is a finitely generated R-module such that $\text{l.dim}_R M =$ gl.dim R.

In case R is a commutative noetherian ring we show gl.dim $A_n(R) =$ gl.dim $R + n$ if R contains the rational numbers; a result also proved in [1, Cor. 2.6] and [7, Th. 24]. If R does not, and m is the maximum of the projective dimensions of cyclic left R-modules that are abelian torsion groups, we prove that gl.dim $A_n(R) =$ gl.dim $R + n$ if $n \leq$ gl.dim $R - m$ and gl.dim $A_n(R) = m + 2n$ if $n \geq$ gl.dim $R - m$; this result has also been proved in [7, Th. 24].

The last section contains examples to illustrate the behavior of gl.dim $A_n(R)$ as a function of n in both the commutative and non-commutative cases.

The work of this paper was carried out in the summer and early fall of 1972, before the untimely and tragic death of the first author on November 2, 1972. Support from the National Science Foundation under grant GP-25600 is gratefully acknowledged.

It is particularly appropriate that this paper, and the one immediately following, appear in a volume dedicated to S. Eilenberg, since many of the notions used in both of them were first introduced in [3].

1. Ore Extensions

Let R be a ring and D a derivation of R. As usual let $S = R[t, D]$ be the Ore extension of R with respect to D, i.e., S is additively the group of polynomials in an indeterminate t with multiplication subject to $tr = rt + D(r)$ for all r in R. We shall extend D to a derivation of S by setting $D(t) = 0$. Thus for all elements s of S we still have $ts = st + D(s)$.

If M is a left (right) S-module, following [10, p. 91] we define an S-module endomorphism $a_l = a_l(M)$ of $S \otimes_R M$ by

$$a_l(s \otimes m) = st \otimes m - s \otimes tm, \qquad s \text{ in } S, \quad m \text{ in } M,$$

where $S \otimes_R M$ is considered as a left S-module via $S \otimes 1$. It is easily verified that

$$0 \longrightarrow S \otimes_R M \xrightarrow{a_l(M)} S \otimes_R M \xrightarrow{b_l(M)} M \longrightarrow 0 \tag{1.1}$$

is an exact sequence of left S-modules where $b_l(s \otimes m) = sm$ [10, p. 91; 11, p. 45]. We shall denote the corresponding maps for right S-modules by a_r and b_r.

Lemma 1.2. Let M be a left S-module and define a right action of S on $M^* = \mathrm{Hom}_R(M, R)$ by

$$(ft)(m) = f(tm) - D(f(m)), \qquad f \text{ in } M^*, \quad m \text{ in } M,$$
$$(fr)(m) = f(m)r.$$

Then M^* becomes a right S-module with this action. Moreover, let $F = F(M)$ be the endomorphism of the abelian group $\mathrm{Hom}_R(M, S)$ defined by $F(g)(m) = tg(m) - g(tm)$. Then the diagram

$$\begin{array}{ccc} M^* \otimes_R S & \xrightarrow{a_r(M^*)} & M^* \otimes_R S \\ \downarrow & & \downarrow \\ \mathrm{Hom}_R(M, S) & \xrightarrow{F(M)} & \mathrm{Hom}_R(M, S) \end{array} \qquad (1.3)$$

where the vertical maps are given by mapping $f \otimes s$ to the element of $\mathrm{Hom}_R(M, S)$ sending m to $f(m)s$ for f in M^*, m in M, and s in S, is commutative.

Proof. The endomorphisms of the abelian group M^* defined by t and any element r of R are readily verified to satisfy $tr = rt + D(r)$. Thus M^* is indeed a right S-module. A routine computation verifies the commutativity of 1.3.

Proposition 1.4. Let R be a left noetherian ring and let M be a left S-module that is finitely generated as an R-module. Then for all $i \geq 1$,

$$\mathrm{Ext}^i_S(M, S) \cong \mathrm{Ext}^{i-1}_R(M, R).$$

Proof. Let $X = \cdots \to X_1 \to X_0 \to M \to 0$ be a projective resolution of M as a left S-module and consider the two complexes $X^* \otimes_R S$ and $\mathrm{Hom}_R(X, S)$. Since the differentiations of X are left S-module homomorphisms, it is routine to verify that $a_r(X^*)$ and $F(X)$ are maps of complexes. Thus 1.3 yields a commutative diagram of complexes, which upon applying the homology functor H in turn, yields a commutative diagram of homology groups

$$\begin{array}{ccc} H(X^* \otimes_R S) & \xrightarrow{H(a_r(X^*))} & H(X^* \otimes_R S) \\ \downarrow & & \downarrow \\ H(\mathrm{Hom}_R(X, S)) & \xrightarrow{H(F(X))} & H(\mathrm{Hom}_R(X, S)) \end{array} \qquad (1.5)$$

We now proceed to identify the homology groups and the maps between them. Since S is a free left R-module, $H(X^* \otimes_R S) \cong H(X^*) \otimes_R S$ [3, IV, Th. 7.2, p. 68] and X is also a projective resolution of M as a left R-module, so that $H(X^*) \cong \mathrm{Ext}_R(M, R)$. Moreover, the differentiations of X^* are easily seen to be right S-module homomorphisms for the right S-module structure of X^* given by Lemma 1.2. Thus, $\mathrm{Ext}_R(M, R)$ also inherits a right S-module structure, and from the definition of $H(a_r(X^*))$ it follows that the isomorphism $H(X^* \otimes_R S) \cong \mathrm{Ext}_R(M, R) \otimes_R S$ carries $H(a_r(X^*))$ to $a_r(\mathrm{Ext}_R(M, R))$.

By [3, VI, Prop. 4.1.3, p. 118],

$$H(\mathrm{Hom}_R(X, S)) \cong H(\mathrm{Hom}_S(S \otimes_R X, S)) \cong \mathrm{Ext}_S(S \otimes_R M, S),$$

and a routine calculation shows that the first isomorphism carries $F(X)$ to $\mathrm{Hom}(a_l(X), S)$. Thus under the isomorphism

$$H(\mathrm{Hom}_R(X, S)) \cong \mathrm{Ext}_S(S \otimes_R M, S),$$

the maps $H(F(X))$ are carried to $\mathrm{Ext}(a_l(M), S)$. Hence we have a commutative diagram

$$\begin{array}{ccc} \mathrm{Ext}_R(M, R) \otimes_R S & \xrightarrow{a_r(\mathrm{Ext}_R(M, R))} & \mathrm{Ext}_R(M, R) \otimes_R S \\ \downarrow & & \downarrow \\ \mathrm{Ext}_S(S \otimes_R M, S) & \xrightarrow{\mathrm{Ext}(a_l, S)} & \mathrm{Ext}_S(S \otimes_R M, S) \end{array} \tag{1.6}$$

where the vertical maps are induced by the sequence of homomorphisms $\mathrm{Hom}_R(X, R) \otimes_R S \to \mathrm{Hom}_R(X, S) \to \mathrm{Hom}_S(S \otimes_R X, S)$. Now, the last map is an isomorphism of complexes; to compute the effect of the first we may replace X by an R-projective resolution Y of M. Since R is left noetherian, Y may be chosen to consist of finitely generated free left R-modules [3, V, Prop. 1.3, p. 78]. But for a finitely generated free left R-module W, the map $\mathrm{Hom}_R(W, R) \otimes_R S \to \mathrm{Hom}_R(W, S)$ given by $(f \otimes s)(w) = f(w)s$ is an isomorphism [3, XI, Th. 3.1, p. 209], so that the vertical maps in 1.6 are isomorphisms.

Now the right analog of 1.1 applied to the top row of 1.6 shows that the maps $\mathrm{Ext}(a_l, S)$ are all injective, so that the usual long exact sequence for Ext arising from 1.1 breaks up into short exact sequences

$$0 \to \mathrm{Ext}_S^i(S \otimes_R M, S) \xrightarrow{\mathrm{Ext}(a_l, S)} \mathrm{Ext}_S^i(S \otimes_R M, S) \to \mathrm{Ext}_S^{i+1}(M, S) \to 0.$$

Since the vertical maps in 1.6 are isomorphisms, and by the right analog of 1.1, the cokernel of $a_r(\mathrm{Ext}_R^i(M, R))$ is $\mathrm{Ext}_R^i(M, R)$, Proposition 1.4 follows.

Following [3] we write $\mathrm{w.dim}_S\, M$ for the flat dimension of a left S-module M, i.e., $\mathrm{w.dim}_S\, M$ is the largest positive integer n such that $\mathrm{Tor}_n^S(N, M) \neq 0$ for some right S-module N. We note that if S is left noetherian and M is a finitely generated left S-module, $\mathrm{w.dim}_S\, M = \mathrm{l.dim}_S\, M$ [3, VI, Ex. 3(b), p. 122] and that S is left (right) noetherian if R is.

Corollary 1.7. Let M be a left S-module.

(a) $\mathrm{w.dim}_R\, M \leq \mathrm{w.dim}_S\, M \leq \mathrm{w.dim}_R\, M + 1$. Thus, $\mathrm{w.gl.dim}\, S \leq \mathrm{w.gl.dim}\, R + 1$.

(b) If $M \neq 0$ is finitely generated as an R-module, R is left noetherian and $\mathrm{l.dim}_R\, M$ is finite, then $\mathrm{l.dim}_S\, M = \mathrm{l.dim}_R\, M + 1$.

Proof. The first inequality in (a) is a consequence of the fact that any flat S-resolution of M is also a flat R-resolution. Combining the long exact sequence for Tor arising from 1.1 with the isomorphisms

$$\mathrm{Tor}^S(N, S \otimes_R M) \cong \mathrm{Tor}^R(N, M)$$

[3, VI, Prop. 4.1.2, p. 117], yields the rest of (a). To prove (b), if $\mathrm{l.dim}_R\, M = i$ then $\mathrm{Ext}_R^i(M, R) \neq 0$ [3, VI, Ex. 9, p. 123] and thus by Proposition 1.4, we have $\mathrm{Ext}_S^{i+1}(M, S) \neq 0$, which yields (b).

Remarks. Corollary 1.7(a) was also proved by Fields [4] using the same methods. In case R is semiprime and also right noetherian, Corollary 1.7(b), proved by entirely different methods, is Theorem 7 of [7]. Corollary 1.7(b) yields alternate proofs of several results in the literature. Thus [8, Prop. 1 and Th. 2] and [6, Lemma 6, Th. 8, Cor. 9], all follow immediately from Corollary 1.7(b). As noted in [6] the results of [8] are only proved if gl.dim R is finite. Finally, we note that combining Corollary 1.7(b) with Theorem 4.2 of [14] yields Theorem 22 of [7].

Combining Corollary 1.7(b) and Theorem 3.8 of [14] yields

Corollary 1.8. Let R be a left and right noetherian ring with gl.dim $R = d < \infty$ and D a derivation of R. Let $S = R[t, D]$ be the ring of differential polynomials over R. Then gl.dim $S = d$ or $d + 1$. The second case occurs if and only if there exists a left S-module M that is a finitely generated left R-module with $\mathrm{l.dim}_R\, M = d$.

2. Weyl Algebras

If R is any ring, we shall write $A_n(R)$ for the Weyl algebra in n variables, i.e., $A_n(R) = R[x_1, x_2, \ldots, x_n, t_1, \ldots, t_n]$ with the x's and t's indeterminates over R commuting with the elements of R, $t_i x_i = x_i t_i + 1$, and $t_i x_j = x_j t_i$, $i \neq j$. It is well known that $A_n(R)$ is left (right) noetherian if R is.

Lemma 2.1. w.gl.dim $R + n \leq$ w.gl.dim $A_n(R) \leq$ w.gl.dim $R + 2n$.

Proof. Since $A_n(R) = A_1(A_{n-1}(R))$, it suffices to prove the lemma for $n = 1$. Now $A_1(R) = R[x][t, d/dx]$ and w.gl.dim $R[x] =$ w.gl.dim $R + 1$ [11, Th. 5.5, p. 45], so that the right-hand inequality follows immediately from Corollary 1.7(a) with $R = R[x]$ and $S = A_1(R)$. Since R is a direct summand of $A_1(R)$ as a R-bimodule, the analog of [11, Th. 5.4, p. 44] for weak dimension shows that w.gl.dim $R \leq$ w.gl.dim $A_1(R)$, so that if w.gl.dim R is infinite, Lemma 2.1 is proved. If w.gl.dim $R = d < \infty$, so that w.gl.dim $R[x] = d + 1$, let M be an $R[x]$-module with $\mathrm{w.dim}_{R[x]}\, M = d + 1$. The exact sequence $0 \to M \to A_1(R) \otimes_{R[x]} M \to \overline{M} \to 0$ shows that $\mathrm{w.dim}_{R[x]}\, A_1(R) \otimes_{R[x]} M = d + 1$, and so the left-hand inequality follows from Corollary 1.7(a).

Lemma 2.1, in case w.gl.dim $R < \infty$, is also an immediate consequence of [6, Prop. 3].

Theorem 2.2. Let R be a left and right noetherian ring of finite global dimension d. Then gl.dim $A_1(R) = d + 2$ if and only if there exists a left $A_1(R)$-module M that is a finitely generated R-module with $\text{l.dim}_R\, M = d$.

Proof. If such an M exists, then applying Corollary 1.7(b) twice (noting $R[x] = R[x, 0]$ and $A_1(R) = R[x][t, d/dx]$) we first find $\text{l.dim}_{R[x]}\, M = d + 1$ and then $\text{l.dim}_{A_1(R)} M = d + 2$. This together with Lemma 2.1 proves the sufficiency of the condition.

To prove necessity, assume gl.dim $A_1(R) = d + 2$. Let M be a finitely generated left $A_1(R)$-module with $\text{l.dim}_{A_1(R)}\, M = d + 2$. Let F denote the family of $A_1(R)$-submodules M' of M with $\text{l.dim}_{A_1(R)}\, M/M' = d + 2$. Since 0 lies in F and $A_1(R)$ is left noetherian, F contains a maximal element, M_0' say. Thus if $M' \supsetneqq M_0'$, we have $\text{l.dim}_{A_1(R)}(M/M') < d + 2$. The long exact sequence for the Ext functor arising from $0 \to M'/M_0' \to M/M_0' \to M/M' \to 0$ then shows $\text{l.dim}_{A_1(R)}(M'/M_0') = d + 2$. By replacing M by M/M_0' we may thus suppose that every nonzero left $A_1(R)$-submodule of M has projective dimension $d + 2$.

By Theorem 3.8 of [14] we may further suppose that M is finitely generated as an $R[x]$-module and that $\text{l.dim}_{R[x]}\, M = d + 1$. Theorem 3.8 of [14] applied once again shows that there is a left $R[x]$-submodule $\overline{M}$ of M that is finitely generated as an R-module. Let $\overline{\overline{M}} = A_1(R)\overline{M} = \sum_0^\infty t^i\overline{M}$. Since $xt^i = t^ix - it^{i-1}$, we have an ascending chain, $\sum_0^k t^i\overline{M}$ of $R[x]$-submodules of $\overline{\overline{M}}$. Since M is a finitely generated left $R[x]$-module and $R[x]$ is left noetherian, we see that $\overline{\overline{M}} = \sum_0^k t^i\overline{M}$ for some natural number k. Thus $\overline{\overline{M}}$ is the desired $A_1(R)$-module since $t^i\overline{M}$ is a finitely generated R-module and by Corollary 1.7(a), applied twice, $d + 2 = \text{l.dim}_{A_1(R)}\, \overline{\overline{M}} \leq \text{l.dim}_R\, \overline{\overline{M}} + 2$.

If R is a division ring of characteristic zero, Theorem 2.2 is [9, Th. 2.4].

In general, it seems quite difficult to apply Theorem 2.2 to compute gl.dim $A_n(R)$ for arbitrary two-sided noetherian rings. However, we can obtain fairly definite answers in case R is commutative. We begin with some preliminaries:

For any ring, let $m = m(R)$ denote the maximum of the weak dimensions of the cyclic left R-modules that are abelian torsion groups.

Lemma 2.3. Let R be a left and right noetherian ring that does not contain the rational numbers.

(a) There are cyclic R-modules that are abelian torsion groups and for every left R-module M that is an abelian torsion group we have $\text{w.dim}_R\, M \leq m(R)$.

(b) $m(A_1(R)) = m(R) + 2$.

Proof. Since R does not contain the rational numbers, the image of some natural number k is not a unit in R. Thus R/Rk is a cyclic left R-module that is an abelian torsion group. Next, let M denote a finitely generated left R-module that is an abelian torsion group. Then there is an exact sequence $0 \to C \to M \to Q \to 0$ with C a cyclic R-module and Q having fewer generators than M. Hence by induction and the long exact sequence for Ext, we have $\text{l.dim}_R\, M \leq m(R)$. Since every module is the direct limit of its finitely generated submodules and the Tor functor commutes with direct limits [3, VI, Prop. 1.3, p. 107], part (a) is proved.

To prove (b), let M be a left $A_1(R)$-module that is an abelian torsion group. Corollary 1.7(a) shows $\text{w.dim}_{A_1(R)}\, M \leq \text{w.dim}_{R[x]}\, M + 1 \leq \text{w.dim}_R\, M + 2 \leq m(R) + 2$. Thus $m(A_1(R)) \leq m(R) + 2$. Now, let M_0 denote a cyclic left R-module that is an abelian torsion group. If k is the additive order of a generator of M_0, then $kM_0 = 0$. Let $\overline{R} = R/Rk$. Since $tx^i = x^i t + ix^{i-1}$ in $A_1(\overline{R})$, it follows that x^k lies in the center of $A_1(\overline{R})$. Let $L = A_1(\overline{R})t + A_1(\overline{R})x^k$. It is readily verified that $A_1(\overline{R})/L$ is isomorphic as an R-bimodule to the free $\overline{R}$-module generated by the images of $1, x, x^2, \ldots, x^{k-1}$ in $A_1(\overline{R})/L$. Hence $M = A_1(\overline{R})/L \otimes_{\overline{R}} M_0$ is a left $A_1(R)$-module that as an R-module is isomorphic to the direct sum of k copies of M_0. Hence by Corollary 1.7(b), $\text{l.dim}_{A_1(R)}\, M = \text{l.dim}_{R[x]}\, M_0 + 1 = \text{l.dim}_R\, M_0 + 2$. Hence $m(A_1(R)) = m(R) + 2$.

Lemma 2.4. Let R be a commutative noetherian ring and let $M \neq 0$ be a left $A_1(R)$-module that is finitely generated as a left R-module, then M is an abelian torsion group.

Proof. Let I be the annihilator of M in $A_1(R)$. Then I is a two sided ideal in $A_1(R)$ that is the kernel of the ring homomorphism $A_1(R) \to \text{End}_R(M)$ given by the left $A_1(R)$-module structure of M. Now since R is commutative and M is a finitely generated R-module, $\text{End}_R(M)$ is also a finitely generated R-module [3, VI, Ex. 2, p. 122] and thus the image of any element of $A_1(R)$ in $\text{End}_R(M)$ is integral over R [12, Th. 12, p. 9]. Hence there exists a monic polynomial f in $R[x]$ such that $f(x)$ lies in I. Now $tf - ft = df/dx$ in $A_1(R)$ so that df/dx also lies in I. Thus all the derivatives of f lie in I, and so if k is the degree of f, the integer $k!$ lies in I or $k!\,M = 0$.

Remark. Theorem 2.2 and Lemma 2.4 show that for a commutative noetherian ring R of global dimension d, containing the rational numbers, $\text{gl.dim}\, A_1(R) = d + 1$, which is Theorem 2.3 of [1].

Lemma 2.5. Let R be a commutative noetherian ring and x an indeterminate over R. Let $R(x)$ denote the ring of fractions of $R[x]$ with regard to the multiplicative set of monic polynomials. Then gl.dim $R(x) =$ gl.dim R.

Proof. The multiplicative set of monic polynomials contains no zero divisors. Moreover, the usual division algorithm holds in $R[x]$ if the divisor is monic. Hence an element of $R(x)$ may be written as $q + r/m$, where q, r, m lie in $R[x]$, m is monic, and degree $r <$ degree m. It is also easy to verify that q is unique. Now, let T denote those elements of $R(x)$ with $q(0) = 0$. Clearly, T is an R-submodule of $R(x)$ and $R(x) = R \oplus T$. Since $R(x)$ is a flat R-module, the analog of [11, Th. 5.4, p. 44] for weak dimensions shows that w.gl.dim $R \leq$ w.gl.dim $R(x)$. Since R, and hence $R(x)$, are noetherian rings, gl.dim $R \leq$ gl.dim $R(x)$. Thus Lemma 2.5 is proved if gl.dim R is infinite.

Next, suppose gl.dim $R = d < \infty$. Since gl.dim $R[x] =$ w.gl.dim $R[x] = d + 1$ and gl.dim $R(x) \leq$ gl.dim $R[x]$ [3, VII, Ex. 10, p. 142], we have gl.dim $R(x) \leq d + 1$. If gl.dim $R(x) =$ w.gl.dim $R(x) = d + 1$, let M be a left $R(x)$-module with $\text{w.dim}_{R(x)} M = d + 1$. By the same reference,

$$\text{w.dim}_{R(x)} M = \text{w.dim}_{R[x]} M = \text{w.gl.dim } R[x].$$

Hence by Theorem 3.8 of [14], the module M contains an $R[x]$-module finitely generated as an R-module. Just as in the proof of Lemma 2.4, this means that this submodule is annihilated by a monic polynomial in x. But this is impossible for an $R(x)$-module, proving the lemma.

Theorem 2.6. Let R be a commutative noetherian ring.

(a) If R contains the rational numbers then

$$\text{gl.dim } A_n(R) = \text{gl.dim } R + n.$$

(b) If R fails to contain the rational numbers, let $m = m(R)$ be the maximum of the projective dimensions of cyclic R-modules that are abelian torsion groups. Then gl.dim $A_n(R) =$ gl.dim $R + n$, if $n \leq$ gl.dim $R - m(R)$, and gl.dim $A_n(R) = m(R) + 2n$, if $n \geq$ gl.dim $R - m(R)$.

Proof. In light of Lemma 2.1, there is nothing to prove if gl.dim R is infinite. We thus assume gl.dim $R = d < \infty$. It is sufficient to show, in light of Lemma 2.1, that

$$m = m(R) \leq d - n \quad \text{implies that} \quad \text{gl.dim } A_n(R) \leq d + n. \tag{2.7}$$

For suppose that this has been done. Then gl.dim $A_{d-m}(R) = 2d - m$. But repeated application of Lemma 2.3(b) shows $m(A_{d-m}(R)) = m + 2(d - m) =$ gl.dim $A_{d-m}(R)$. Again, Lemmas 2.1 and 2.3(b) show gl.dim $A_k(R') =$ gl.dim $R' + 2k$ in case $m(R') =$ gl.dim R' and R' is left and right noetherian.

Then if $n = d - m + k$ say, and gl.dim $A_{d-m}(R) = 2d - m$, we have gl.dim $A_n(R) =$ gl.dim $A_k(A_{d-m}(R)) = 2d - m + 2k = m + 2n$. Of course in case (a), we take $m(R) = -\infty$.

We now proceed to prove (2.7) by induction on n. Clearly (2.7) is true if $n = 0$. Now let $T(x_n) = A_{n-1}(R(x_n))[t_n, d/dx_n]$. Let M be a left $R(x_n)$-module that is an abelian torsion group. Then

$$\begin{aligned} \text{w.dim}_{R(x_n)} M = \text{w.dim}_{R[x_n]} M \leq \text{w.dim}_R M + 1 &\leq m(R) + 1 \\ &\leq d - n + 1 = d - (n-1). \end{aligned}$$

Thus by Lemma 2.3(a), $m(R(x_n)) \leq d - (n-1) =$ gl.dim $R(x_n) - (n-1)$ by Lemma 2.5. Hence by induction, gl.dim $A_{n-1}(R(x_n)) \leq$ gl.dim $R(x_n) + (n-1)$, which by Lemma 2.5 is $\leq d + (n-1)$. Now $T(x_n)$ can be viewed as the ring of differential polynomials of $A_{n-1}(R(x_n))$ with regard to the derivation d/dx_n. Thus by Corollary 1.7(a), gl.dim $T(x_n) \leq d + n$.

Suppose $d' =$ gl.dim $A_n(R) > d + n$. Let N be a finitely generated left $A_n(R)$-module with $\text{l.dim}_{A_n(R)} N = d'$. Just as in the proof of Theorem 2.2, we may suppose that all nonzero $A_n(R)$-submodules of N have projective dimension d' also. Now it is clear that $T(x_n) \cong R(x_n) \otimes_{R[x_n]} A_n(R)$ as a right $A_n(R)$-module, and similarly on the left. Since $R(x_n)$ is flat as a $R[x_n]$-module, we see that $T(x_n)$ is flat as both a right and left $A_n(R)$-module. Therefore by [3, VI, Ex. 10, p. 123],

$$\text{w.dim}_{A_n(R)} T(x_n) \otimes_{A_n(R)} N \leq \text{w.dim}_{T(x_n)} T(x_n) \otimes_{A_n(R)} N \leq d + n.$$

Now let

$$\begin{aligned} \bar{N} = T(x_1) \otimes_{A_n(R)} N \oplus \cdots \oplus T(x_n) \otimes_{A_n(R)} N \\ \oplus T(t_1) \otimes_{A_n(R)} N \oplus \cdots \oplus T(t_n) \otimes_{A_n(R)} N. \end{aligned}$$

Since $T(x_i)$ and $T(t_j)$ behave like $T(x_n)$ it follows that $\text{w.dim}_{A_n(R)} \bar{N} \leq d + n < d'$. Now $T(x_n) \otimes_{A_n(R)} N \cong R(x_n) \otimes_{R[x_n]} N$. Hence the kernel of the natural map $N \to T(x_n) \otimes_{A_n(R)} N$ consists of the elements of N annihilated by monic polynomials in x_n. Hence if z lies in Ker$(N \to \bar{N})$, z must be annihilated by monic polynomials in each of $x_1, \ldots, x_n, t_1, \ldots, t_n$. It is then easily verified than $\bar{\bar{N}} = A_n(R)z$ is finitely generated over R and thus, by Lemma 2.4, an abelian torsion group. Now if $\bar{\bar{N}} \neq 0$ then $\text{l.dim}_{A_n(R)} \bar{\bar{N}} = d' > d + n$. But repeated application of Corollary 1.7(b) shows that $\text{l.dim}_{A_n(R)} \bar{\bar{N}} = \text{l.dim}_R \bar{\bar{N}} + 2n$. Thus $m(R) + 2n > d + n$ or, $m(R) > d - n$, contradicting the hypothesis of 2.7. Hence $\bar{\bar{N}} = 0$ and $N \to \bar{N}$ is an injection. But then it follows from the long exact sequence for the Tor functor applied to the exact sequence $0 \to N \to \bar{N} \to \bar{N}/N \to 0$, that w.dim $\bar{N}/N > d'$, which is impossible. Thus 2.7, and with it Theorem 2.6, is proved.

Corollary 2.8. If R is a noetherian commutative ring of finite global dimension d, gl.dim $A_1(R) = d + 2$ if and only if R has a residue class field F of finite characteristic with $\text{l.dim}_R F = d$.

Proof. From Theorem 2.6 it is clear that gl.dim $A_1(R) = d + 2$ can only occur if $m(R) = d$. But it is well known that $d = \sup$ gl.dim R_P, where P runs through the maximal ideals of R and gl.dim $R_P = \text{l.dim}_R R/P$.

Remarks. In case R is a field of characteristic zero, Theorem 2.6(a) is due to Roos [13]. We gratefully acknowledge making use of his ideas in our proof of Theorem 2.6. Theorem 2.6(a) has also been proved in [1, Cor. 2.6] by methods somewhat similar to our own. All of Theorem 2.6 has been proved by different methods in [7, Th. 24].

It is worth noting that the ring $T(x_1) \oplus T(x_2) \oplus \cdots \oplus T(x_n) \oplus T(t_1) \oplus T(t_2) \oplus \cdots \oplus T(t_n)$ need not be faithfully flat over $A_n(R)$, so that in order to prove 2.7 it is not possible to simply apply [2, Lemma 2.b.4]. To obtain an example let R denote a field of characteristic 2 and let $n = 1$. Then x_1^2 and t_1^2 are central elements of $A_1(R)$ that generate a proper ideal I; in fact $A_1(R)/I \cong R_2$, the ring of 2×2 matrices over R. However, since x_1^2 is a unit in $T(x_1)$ and t_1^2 is a unit in $T(t_1)$, it is clear that $I(T(x_1) \oplus T(t_1)) = T(x_1) \oplus T(t_1)$ so that $T(x_1) \oplus T(t_1)$ is not faithfully flat over $A_1(R)$.

3. Examples

Let R be any ring of finite left global dimension. Setting $\Delta_n(R) =$ l.gl.dim $A_n(R) -$ l.gl.dim $A_{n-1}(R)$, we have by Lemma 2.1 that $\Delta_n(R) = 1$ or 2. By Theorem 2.6 there are only two possible patterns in case R is a commutative noetherian ring: If R contains the field of rational numbers Q, $\Delta_n(R) = 1$ for all n. Otherwise, there is a natural number k such that $\Delta_n(R) = 1$ when $n \leq k$ and $\Delta_n(R) = 2$ when $n > k$. An example of this second pattern is furnished by $R = Q[w_1, \ldots, w_k] \oplus Z/2Z$, where $w_1, \ldots, w_k$ are indeterminates over Q. Here gl.dim $R = k$ and $m(R) = 0$. Since $A_n(R) = A_n(Q[w_1, \ldots, w_k]) \oplus A_n(Z/2Z)$, Theorem 2.6 shows that gl.dim $A_n(R) = \text{Max}(n + k, 2n)$. Thus $\Delta_n(R) = 1$ for $n \leq k$ and $\Delta_n(R) = 2$ for $n > k$.

In the noncommutative case the situation can be more complicated as is illustrated by the following examples:

Example 3.1. By [5, p. 503], $A_s(Q)$ has a division ring of quotients which we denote by R. We relabel the x's and t's in R as u's and v's. If $n \leq s$ the additive group of R has a structure as left $A_n(R)$-module given by the usual left R structure and by $x_i r = rv_i$, $t_j r = ru_j$. Clearly, Corollary 1.7(b) implies that $\text{l.dim}_{A_n(R)} R = 2n$ for $n \leq s$. Hence $\Delta_n(R) = 2$ if $n \leq s$.

On the other hand, it is clear that for all n, $A_n(R)$ can be obtained from $A_{n+s}(Q)$ as a ring of fractions. Thus $A_n(R)$ is flat as both a right and left $A_{n+s}(Q)$-module and $A_n(R) \otimes_{A_{n+s}(Q)} A_n(R) \cong A_n(R)$. Hence by [2, Lemma 2.b.2], gl.dim $A_n(R) \leq$ gl.dim $A_{n+s}(Q) = n + s$. Therefore, if $n > s$, we have $\Delta_n(R) = 1$.

Example 3.2. Let R be the division ring defined in Example 3.1. Then $A_n(R[w_1, \ldots, w_k]) = A_n(R)[w_1, \ldots, w_k]$ so that gl.dim $A_n(R[w_1, \ldots, w_k)] =$ gl.dim $A_n(R) + k$. Hence if $S = R[w_1, \ldots, w_k] \oplus Z/2Z$, it is readily seen that $\Delta_n(S) = 2$ when $n \leq s$ or $n > s + k$ and $\Delta_n(S) = 1$ otherwise.

Example 3.3. Again let R be the division ring of Example 3.1 with $s > 1$ and let $R' = R \oplus Q[w_1, \ldots, w_k]$ with $k < s$. Then

$$\text{gl.dim } A_n(R') = \begin{cases} n + k, & 0 \leq n \leq k \\ 2n, & k \leq n \leq s \\ n + s, & s \leq n. \end{cases}$$

Thus $\Delta_n(R') = 1,\ 0 \leq n \leq k;\ \Delta_n(R') = 2,\ k < n \leq s;\ \Delta_n(R') = 1,\ s < n$. This example is due, independently, to K. R. Goodearl and J. T. Stafford.

We have thus exhibited the following possible sequences for $\Delta_n(R)$:

$$1, 1, 1, \ldots$$
$$1, 1, 1, \ldots, 1, 2, 2, \ldots$$
$$2, 2, 2, \ldots, 2, 1, 1, \ldots$$
$$2, 2, 2, \ldots, 2, 1, 1, \ldots, 1, 2, 2, \ldots$$
$$1, 1, 1, \ldots, 1, 2, 2, \ldots, 2, 1, 1, \ldots.$$

Indeed, Goodearl and Stafford have independently pointed out that one can construct a ring R such that $\Delta_n(R)$ is any prescribed sequences of ones and twos, provided that it is ultimately constant.

REFERENCES

[1] S. M. Bhatwadekar, On the global dimension of Ore extensions, *Nagoya Math. J.* **50** (1973), 217–225.

[2] J. E. Björk, The global homological dimension of some algebras of differential operators, *Invent Math.* **17** (1972), 67–78.

[3] H. Cartan and S. Eilenberg, "Homological Algebra." Princeton Univ. Press, Princeton, New Jersey, 1956.

[4] K. Fields, On the global dimension of skew polynomial rings—an addendum, *J. Algebra* **14** (1970), 528–530.

[5] I. M. Gelfand and A. A. Kirillov, Sur les corps liés aux algèbres enveloppantes des algèbres de Lie, *Inst. Hautes Etudes Sc. Publ. Math.* **31** (1966), 509–523.

[6] K. R. Goodearl, Global dimension of differential operator rings, *Proc. Amer. Math. Soc.* **45** (1974), 315–322.
[7] K. R. Goodearl, Global dimensions of differential operator rings, II, *Trans. Amer. Math. Soc.* **209** (1975), 65–85.
[8] N. S. Gopalakrishnan and R. Sridharan, Homological dimension of Ore extensions, *Pacific J. Math.* **19** (1968), 67–75.
[9] R. Hart, A note on the tensor products of algebras, *J. Algebra* **21** (1972), 422–427.
[10] G. Hochschild, A note on relative homological dimension, *Nagoya Math. J.* **13** (1958), 89–94.
[11] I. Kaplansky, Commutative rings, "Queen Mary College Lecture Notes." Queen Mary College, London, 1966.
[12] I. Kaplansky, "Commutative Rings." Allyn & Bacon, Boston, Massachusetts, 1970.
[13] J. E. Roos, Détermination de la dimension homologique globale des algèbres de Weyl, *C.R. Acad. Sci. Paris, Ser. A* **274** (1972), 23–26.
[14] A. Rosenberg and J. T. Stafford, Global dimension of Ore extensions (this volume, Ch. 14).

Primary AMS 16A60; Secondary 16A72

DEPARTMENT OF MATHEMATICS
CORNELL UNIVERSITY
ITHACA, NEW YORK

Global Dimension of Ore Extensions

ALEX ROSENBERG and J. T. STAFFORD

1. Introduction

Let R be a ring with derivation D and $S = R[t]$, $tr = rt + D(r)$ for all r in R, the corresponding Ore extension. Several authors, among them [4] and [6], have noted the inequality

$$\text{l.gl.dim } S \leq \text{l.gl.dim } R + 1.$$

In this chapter we show that if R is left and right noetherian and of finite left global dimension, a necessary condition for equality to hold is the existence of a left S-module M that is finitely generated as an R-module and with $\text{l.dim}_R\, M = \text{l.gl.dim } R$. Combining the result of this chapter with Corollary 1.7(b) of [9] shows that the given condition is sufficient as well as necessary. We also show that in case R is commutative, our result implies part of Theorem 22 of [5].

This chapter grew out of a proof of Theorem 3.8 for the case $D = 0$ due to the late G. S. Rinehart and the first author. Rinehart's contribution is hereby gratefully acknowledged. The first author also thanks the National Science Foundation of the United States for its support under grants GP-25600 and GP-40773X. The second author wishes to thank the Science Research Council of Great Britain for its support and also gratefully acknowledges the help and encouragement of J. C. Robson.

2. Tor Functor and Intersections

In this section we prove the formula

$$\mathrm{Tor}_d^R(N, \cap M_i) = \cap \mathrm{Tor}_d^R(N, M_i) \tag{2.1}$$

for a right noetherian ring R with w.gl.dim $R = d$, a right R-module N, and a family $\{M_i\}$ of submodules of a left module M. We begin with

Lemma 2.2. Let R denote a right noetherian ring with w.gl.dim $R = d < \infty$, N a fixed arbitrary right R-module and F the functor defined by $F(X) = \mathrm{Tor}_d^R(N, X)$ for any left R-module X. Then

(i) F is left exact (as a functor of both variables). In particular, if $X' \subset X$ are left R-modules, the map $F(X') \to F(X)$ is injective, and we shall consistently identify $F(X')$ with its image in $F(X)$.

(ii) For any family $\{X_i\}$ of left R-modules, the map $F(\Pi X_i) \to \Pi F(X_i)$ arising from the projection maps is monic.

Proof. Part (i) is immediate from the definition of w.gl.dim and the appropriate long exact homology sequence. As for Part (ii), by [3, II, Ex. 2, pp. 31–32], for any right R-module Y, there is a natural homomorphism $Y \otimes_R \Pi X_i \to \Pi(Y \otimes_R X_i)$ arising from the projection maps. This is an isomorphism if Y is finitely generated over R. If Y is a projective resolution of N, we obtain homomorphisms

$$\varphi_n\colon \mathrm{Tor}_n^R(N, \Pi X_i) \to \Pi\, \mathrm{Tor}_n^R(N, X_i)$$

[3, V, Prop. 9.3, p. 98]. Since R is right noetherian, whenever N is finitely generated there is a projective resolution of N consisting of finitely generated R-modules [3, V, Prop. 1.3, p. 78]. Thus φ_n is an isomorphism if N is finitely generated. Now for any right R-module N and submodule N' there is a commutative diagram

$$\begin{array}{ccc}
F(\Pi X_i) = \mathrm{Tor}_d^R(N, \Pi X_i) & \xrightarrow{\varphi_d} & \Pi\, \mathrm{Tor}_d^R(N, X_i) = \Pi F(X_i) \\
\uparrow & & \uparrow \\
\mathrm{Tor}_d^R(N', \Pi X_i) & \xrightarrow{\varphi_d'} & \Pi\, \mathrm{Tor}_d^R(N', X_i)
\end{array}$$

where, by Part (i), the vertical maps are monomorphisms induced by the inclusion $N' \to N$. By [3, VI, Ex. 17, p. 125], $F(\Pi X_i)$ is the union of the subgroups of $\mathrm{Tor}_d^R(N', \Pi X_i)$, where N' runs through the finitely generated submodules of N. Hence an element of Ker φ_d would have to lie in $\mathrm{Tor}_d^R(N', \Pi X_i)$ for some finitely generated submodule N'. Since φ_d' and the two vertical maps are injective, a simple diagram chase shows that Ker $\varphi_d = 0$.

Lemma 2.3. Let F denote a functor from the category of left R-modules to the category of abelian groups satisfying Parts (i) and (ii) of Lemma 2.2. Then for any R-module M and a family $\{M_i\}$ of submodules of M,

$$\cap F(M_i) = F(\cap M_i).$$

Proof. Denote the projections $\Pi M/M_i \to M/M_i$ and $M \to M/M_i$ by p_i and q_i, respectively. The commutative diagram

$$\begin{array}{ccccccc} 0 & \longrightarrow & \cap M_i & \longrightarrow & M & \xrightarrow{\Pi q_i} & \Pi M/M_i \\ & & & & \| & & \downarrow{\scriptstyle p_j} \\ & & & & M & \xrightarrow{q_j} & M/M_j \end{array}$$

with exact top row, yields in view of the left exactness of F, a commutative diagram with exact rows:

$$\begin{array}{ccccccc} 0 & \longrightarrow & F(\cap M_i) & \longrightarrow & F(M) & \xrightarrow{F(\Pi q_i)} & F(\Pi(M/M_i)) \\ & & & & \| & & \downarrow{\scriptstyle \Pi F(p_i)} \\ 0 & \longrightarrow & \cap F(M_i) & \longrightarrow & F(M) & \xrightarrow{\Pi F(q_i)} & \Pi F(M/M_i) \end{array}$$

Now, if A, B, C denote abelian groups and $f\colon A \to B$, $g\colon A \to C$, $s\colon B \to C$ homomorphisms such that $sf = g$, it is clear that $\operatorname{Ker} f \subset \operatorname{Ker} g$. If in addition s is monic, it follows that $\operatorname{Ker} f = \operatorname{Ker} g$. Hence $F(\cap M_i) = \cap F(M_i)$.

It is clear that Lemmas 2.2 and 2.3 prove (2.1).

3. Ore Extensions

Let R be a ring with derivation D. As usual $S = R[t]$ is the Ore extension of R with respect to D, i.e., S is additively the group of polynomials in an indeterminate t with multiplication subject to $tr = rt + D(r)$ for all r in R. If w.gl.dim $R = d < \infty$, we investigate the consequence of w.gl.dim $S = d + 1$.

We recall that for any left S-module M there is an exact sequence of left S-modules

$$0 \longrightarrow S \otimes_R M \xrightarrow{a_l(M)} S \otimes_R M \xrightarrow{b_l(M)} M \longrightarrow 0, \tag{3.1}$$

where $S \otimes_R M$ is treated as a left S-module via $S \otimes 1$ and a_l and b_l are defined by

$$a_l(s \otimes m) = st \otimes m - s \otimes tm, \qquad b_l(s \otimes m) = sm$$

[7, p. 91; 8, Proof of Th. 6, pp. 174–6]. Let N denote a right S-module and Z an S-projective resolution of N. It is easily checked that $\alpha(z \otimes m) = zt \otimes m - z \otimes tm$ defines an endomorphism of $Z \otimes_R M$ and that the diagram

$$\begin{array}{ccc} Z \otimes_S S \otimes_R M & \xrightarrow{1 \otimes a_l(M)} & Z \otimes_S S \otimes_R M \\ \downarrow & & \downarrow \\ Z \otimes_R M & \xrightarrow{\alpha} & Z \otimes_R M \end{array}$$

where the vertical maps are defined by $z \otimes s \otimes m \to zs \otimes m$, is commutative. Moreover, since the differentiations of Z are right S-homomorphisms, α is a map of complexes. Furthermore, since S is R-free, an S-projective resolution is also an R-projective resolution, so that passing to homology yields the commutative diagram

$$\begin{array}{ccc} \mathrm{Tor}^S(N, S \otimes_R M) & \xrightarrow{\mathrm{Tor}^S(N,\, a_l(M))} & \mathrm{Tor}^S(N, S \otimes_R M) \\ \downarrow & & \downarrow \\ \mathrm{Tor}^R(N, M) & \xrightarrow{\mathrm{Tor}^R(\alpha)} & \mathrm{Tor}^R(N, M) \end{array} \tag{3.2}$$

where the vertical maps are isomorphisms by [3, VI, Prop. 4.1.1, p. 117].

Lemma 3.3. Let R denote a ring with w.gl.dim $R = d < \infty$. Then V, the kernel of the endomorphism $\mathrm{Tor}_d^R(\alpha)$, is isomorphic to $\mathrm{Tor}_{d+1}^S(N, M)$.

Proof. The long exact sequence of Tor^S induced by (3.1) together with the isomorphism $\mathrm{Tor}^S(N, S \otimes_R M) \cong \mathrm{Tor}^R(N, M)$, shows that

$$\mathrm{Ker}(\mathrm{Tor}_d^S(N, a_l(M))$$

is isomorphic to $\mathrm{Tor}_{d+1}^S(N, M)$. The lemma then follows from (3.2).

Next, let M' denote an R-submodule of the S-module M. For $i = 0, 1, 2, \ldots$, set

$$K_i = \{x \text{ in } M' \mid t^j x \text{ in } M',\ j = 0, 1, 2, \ldots, i\}.$$

Clearly, $K_i \supset K_{i+1}$. As is well known, for all r in R,

$$t^j r = rt^j + jD(r)t^{j-1} + \binom{j}{2} D^2(r)t^{j-2} + \cdots; \tag{3.4}$$

thus the K_i are R-submodules of M'. Let π_i: $K_i \to M/M'$ be defined by $\pi_i(x) = t^{i+1}x + M'$. Using (3.4), it is readily seen that π_i is an R-homomorphism so that

$$0 \longrightarrow K_{i+1} \longrightarrow K_i \xrightarrow{\pi_i} M/M' \tag{3.5}$$

is an exact sequence of R-modules.

Lemma 3.6. Let R be a ring with w.gl.dim $R = d < \infty$. Then, using the convention of Lemma 2.2(i), for any v in $V \cap \mathrm{Tor}_d^R(N, K_i)$, we have $\mathrm{Tor}_d^R(N, \pi_i)(v) = 0$.

Proof. As before, let Z be an S-projective resolution of N. Let f be a cocycle representing the cohomology class v. By Lemma 2.2(i), we may assume

$$f = \sum z_h \otimes k_h, \qquad \text{with } k_h \text{ in } K_i.$$

Since v is in V, the cohomology class of $\alpha(f)$ is trivial, i.e., if δ denotes the differentiation of Z, there exists an element w of $Z \otimes_R M$ such that

$$\alpha(f) = \sum z_h t \otimes k_h - \sum z_h \otimes tk_h = (\delta \otimes 1)(w).$$

Now, a straightforward induction proves that

$$\alpha^n(z \otimes k) = zt^n \otimes k - nzt^{n-1} \otimes tk + \binom{n}{2} zt^{n-2} \otimes t^2k - \cdots \pm z \otimes t^n k.$$

Thus, modulo $Z \otimes M'$ we have

$$\sum_h z_h \otimes t^{i+1} k_n \equiv \pm\alpha^{i+1}(f) = \pm\alpha^i(\delta \otimes 1)w = \pm(\delta \otimes 1)(\alpha^i w).$$

But $\mathrm{Tor}_d^R(N, \pi_i)(v)$ is represented in $Z \otimes M/M'$ by $\sum z_h \otimes t^{i+1} k_h + M'$, and thus is a coboundary in $Z \otimes M/M'$, which proves the lemma.

For any R-submodule M' of M we let $K(M') = K = \cap K_i$, the largest S-submodule contained in M'.

Lemma 3.7. If R is right noetherian, w.gl.dim $R = d < \infty$, and M' is a submodule of M such that $\mathrm{w.dim}_R\, K < d$, then $V \cap \mathrm{Tor}_d^R(N, M') = 0$.

Proof. The hypothesis and (2.1) imply that $0 = \mathrm{Tor}_d^R(N, K) = \cap \mathrm{Tor}_d^R(N, K_i)$. Thus, using the convention of Lemma 2.2(i), the $\mathrm{Tor}_d^R(N, K_i)$ are a descending chain of subgroups of $\mathrm{Tor}_d^R(N, M')$, intersecting in 0. Hence if v is a nonzero element of $V \cap \mathrm{Tor}_d^R(N, M')$, there is a natural number i such that v lies in $\mathrm{Tor}_d^R(N, K_i)$ but does not lie in $\mathrm{Tor}_d^R(N, K_{i+1})$. Now by (3.5), Lemma 2.2(i), and [3, II, Prop. 4.3a(b), p. 25],

$$0 \longrightarrow \mathrm{Tor}_d^R(N, K_{i+1}) \longrightarrow \mathrm{Tor}_d^R(N, K_i) \xrightarrow{\mathrm{Tor}_d^R(N, \pi_i)} \mathrm{Tor}_d^R(N, M/M')$$

is exact. But then, $\mathrm{Tor}_d^R(N, \pi_i)(v) \neq 0$, contradicting Lemma 3.6. Hence $V \cap \mathrm{Tor}_d^R(N, M') = 0$.

Theorem 3.8. Let R be a left and right noetherian ring, gl.dim. $R =$ w.gl.dim $R = d < \infty$, and $S = R[t]$ the Ore extension of R with regard to a derivation D. If M is a left S-module with $\text{w.dim}_S M = d + 1$, then M contains an S-submodule M_0 that is a finitely generated R-module and $\dim_R M_0 = \text{w.dim}_R M_0 = d$.

Proof. By hypothesis, there exists a right S-module N such that $\text{Tor}_{d+1}^S(N, M) \neq 0$. Let M' be an arbitrary finitely generated R-submodule of M. Since R is left noetherian, $K(M') = K$ as defined above is also a finitely generated R-module. Thus if the theorem were false, $\text{w.dim}_R K(M')$ would be less than d and so Lemma 3.7 would force $V \cap \text{Tor}_d^R(N, M') = 0$ for all finitely generated R-submodules of M. Using the convention of Lemma 2.2(i) and [3, VI, Ex. 17, p. 125], we have $\text{Tor}_d^R(N, M) = \cup \text{Tor}_d^R(N, M')$, M' running through all the finitely generated R submodules of M. Thus V would be zero, contradicting Lemma 3.3.

4. The Commutative Case

If R is commutative, we can deduce part of Theorem 22 of [5] from Theorem 3.8.

Lemma 4.1. Let R denote a commutative noetherian ring with $\text{gl.dim } R = d < \infty$, $\mathfrak{M}$ the set of maximal ideals of R, and M a finitely generated left R-module with $\dim_R M = d$. Then there exists an element $\mathfrak{m}_0$ of $\mathfrak{M}$ such that $\dim_R R/\mathfrak{m}_0 = d$ and a nonzero element x of M such that, $\mathfrak{m}_0 x = 0$.

Proof. By [3, VII, Ex. 9–11, pp. 141–2],

$$d = \sup_{\mathfrak{m} \text{ in } \mathfrak{M}} \dim_{R_\mathfrak{m}} M_\mathfrak{m} = \sup_{\mathfrak{m} \text{ in } \mathfrak{M}} \text{gl.dim } R_\mathfrak{m},$$

and for any $R_\mathfrak{m}$-module P, we have $\dim_{R_\mathfrak{m}} P = \dim_R P$. Hence there is an element $\mathfrak{m}_0$ in $\mathfrak{M}$ such that $\text{gl.dim } R_{\mathfrak{m}_0} = \dim_{R_{\mathfrak{m}_0}} M_{\mathfrak{m}_0} = d$. But then also

$$\dim_R R/\mathfrak{m}_0 = \dim_{R_{\mathfrak{m}_0}} R_{\mathfrak{m}_0}/R_{\mathfrak{m}_0}\mathfrak{m}_0 = d$$

[8, Th. 12, 13, pp. 182–3].

Now by [1, Prop. 2.2], there is a nonzero element x/s in $M_{\mathfrak{m}_0}$, with s in $R - \mathfrak{m}_0$, whose annihilator is $R_{\mathfrak{m}_0}\mathfrak{m}_0$. Thus, if $\mathfrak{m}_0 = (r_1, \ldots, r_n)$, there are elements $s_1, \ldots, s_n$ in $R - \mathfrak{m}_0$ such that $s_i r_i x = 0$. Since $R - \mathfrak{m}_0$ is a multiplicatively closed set, $0 \neq s_1 s_2 \cdots s_n$ lies in $R - \mathfrak{m}_0$ and, clearly, $\mathfrak{m}_0(s_1 s_2 \cdots s_n x) = 0$. Since $x/s \neq 0$ in $M_{\mathfrak{m}_0}$, the element $s_1 s_2 \cdots s_n x$ is nonzero in M, and so Lemma 4.1 is proven.

Theorem 4.2 (cf. [5, Th. 22]). Let R be a commutative noetherian ring, $\text{gl.dim } R = d < \infty$, and $S = R[t]$ the Ore extension of R with regard to a derivation D. If $\text{gl.dim } S = d + 1$, then there exists a maximal ideal

$\mathfrak{m}_0$ of R such that $\dim_R R/\mathfrak{m}_0 = d$, and either $D(\mathfrak{m}_0) \subset \mathfrak{m}_0$ or there exists a rational prime contained in $\mathfrak{m}_0$.

Proof. By Theorem 3.8, there exists a left S-module M that is finitely generated as an R-module with $\text{w.dim}_R\, M = \dim_R\, M = d$. Let $\mathfrak{m}_0$ be a maximal ideal of R satisfying the conditions of Lemma 4.1 and x an element of M such that $\text{Ann}_R(x) = \mathfrak{m}_0$. Now set

$$M' = \{y \text{ in } M \,|\, \text{Ann}_R(y) \supset \mathfrak{m}_0^{n(y)}\}.$$

It is then easily verified that M' is an R-submodule of M, which is nonzero because it contains x.

Now if for any y in M we have $r_1 y = r_2 y = 0$, then $(r_1 r_2)ty = (r_1 t)r_2 y - r_1 D(r_2)y = 0$, i.e., $\text{Ann}_R(ty) \supset (\text{Ann}_R\, y)^2$. It follows that M' is also an S-module.

Since M' is a finitely generated R-module, $\text{Ann}_R(M') \supset \mathfrak{m}_0{}^n$ for some n. Thus $\text{Ann}_R(M')$ is a primary ideal of R with radical $\mathfrak{m}_0$ [10, III, Cor. 1, p. 153]. Because M' is an S-module, it is clear that if r lies in $\text{Ann}_R(M')$, so does $D(r) = tr - rt$, i.e., $D(\text{Ann}_R\, M') \subset \text{Ann}_R\, M'$.

Now for any r in $\mathfrak{m}_0$, there is a natural number l with r^l in $\text{Ann}_R(M')$ but r^{l-1} not in $\text{Ann}_R(M')$. Hence $D(r^l) = \{l\, D(r)\}r^{l-1}$ lies in $\text{Ann}_R(M')$. Therefore, $l\, D(r)$ lies in the prime ideal $\mathfrak{m}_0$. Thus if $\mathfrak{m}_0$ does not contain a rational prime, $D(r)$ is in $\mathfrak{m}_0$ for all r in $\mathfrak{m}_0$, which completes the proof.

We note that [2, Th. 1.1] is an immediate consequence of this last result and the general inequality $\text{gl.dim}\, S \leq \text{gl.gim}\, R + 1$ noted by several authors.

REFERENCES

[1] M. Auslander and D. Buchsbaum, Homological dimension in local rings, *Trans. Amer. Math. Soc.* **85** (1957), 390–405.

[2] S. M. Bhatwadekar, On the global dimension of Ore extensions, *Nagoya Math. J.* **50** (1973), 217–225.

[3] H. Cartan and S. Eilenberg, "Homological Algebra." Princeton Univ. Press, Princeton, New Jersey, 1956.

[4] K. Fields, On the global dimension of skew polynomial rings—an addendum, *J. Algebra* **14** (1970), 528–530.

[5] K. R. Goodearl, Global dimension of differential operator rings, II, *Trans. Amer. Math. Soc.* **209** (1975), 65–85.

[6] N. S. Gopalakrishnan and R. Sridharan, Homological dimension of Ore extensions, *Pacific J. Math.* **19** (1968), 67–75.

[7] G. Hochschild, A note on relative homological dimension, *Nagoya Math. J.* **13** (1958), 89–94.

[8] I. Kaplansky, "Fields and Rings," Chicago Lectures in Mathematics. Univ. of Chicago Press, Chicago, Illinois, 1969.

[9] G. S. Rinehart and A. Rosenberg, The global dimension of Ore extensions and Weyl algebras (this volume, Ch. 13).

[10] O. Zariski and P. Samuel, "Commutative Algebra." Van Nostrand-Reinhold, Princeton, New Jersey, 1958.

Primary AMS 16A60; Secondary 16A72

Alex Rosenberg
DEPARTMENT OF MATHEMATICS
CORNELL UNIVERSITY
ITHACA, NEW YORK

J. T. Stafford
SCHOOL OF MATHEMATICS
THE UNIVERSITY OF LEEDS
LEEDS, ENGLAND

On the Spectrum of a Ringed Topos

MYLES TIERNEY

This paper is dedicated to Samuel Eilenberg, whose insight and teaching, both mathematical and personal, have meant a great deal to me.

One of the original reasons for introducing the axioms of an elementary topos, though by no means the only one, was to do topos theory intrinsically. That is, we wanted to be able to carry out basic arguments and constructions in a topos without constantly having to choose a site of definition. As an illustration, Lawvere and I, in 1970, chose to give a presentation, in our setting, of the spectrum of a ringed topos. This could then be compared with that of Hakim [3] in the Grothendieck case. The resulting very pretty construction, due primarily to Lawvere and recalled below, turned out to be incorrect. This was shown by an interesting calculation of Joyal in 1972, which dramatically exhibited the scarcity of "primes" in the general case. Joyal himself had an alternative construction, described very sketchily in [6], and I also found one in 1973 based on the notion of "radical ideal." The purpose of this paper is to give a brief account of these developments, which I think are instructive, as well as to give a completely new construction, which seems, from the point of view of the universal property involved, to be about as simple as it could possibly be. The technique involved also has wide applications to other problems, as I hope will become clear below.

Before approaching the problem of the spectrum, we need to establish, in an arbitrary ringed topos, some basic results of commutative algebra. This is done in the following section.

1. Localization, Primes, Local Rings, Etc.

Let $(\mathbf{E}, A)$ be a ringed topos, i.e., $\mathbf{E}$ is a topos (elementary [8]), and A is a commutative ring in $\mathbf{E}$. Previously the terminology was that A is a ring *object* in $\mathbf{E}$. The word "object," however, seems superfluous these days—one does not, after all, use it when $\mathbf{E} = \mathbf{S}$, the category of sets. For this paper only, we denote the initial object of $\mathbf{E}$ by $\varnothing$, to distinguish it from the zero, $1 \xrightarrow{0} A$, of A. The multiplicative identity of A is written $1 \xrightarrow{e} A$, to distinguish it from the terminal object 1.

We define the subobject $U(A) \rightarrowtail A$ of *units of* A by interpreting the formula $\exists a'(a'a = e)$. For readers unaccustomed to this procedure, this means the following: First, using the multiplication on A, define the pullback

$$\begin{array}{ccc} I(A) & \longrightarrow & 1 \\ \downarrow & & \downarrow e \\ A \times A & \xrightarrow{\ \cdot\ } & A \end{array}$$

In set theoretical terms,

$$I(A) = \{(a', a) \mid a'a = e\}.$$

Now take the image:

$$\begin{array}{ccc} I(A) & \twoheadrightarrow & U(A) \\ \downarrow & & \downarrow \\ A \times A & \xrightarrow{\pi_2} & A \end{array}$$

In fact, as in $\mathbf{S}$, the inverse is unique when it exists, so that $I(A) \twoheadrightarrow U(A)$ is an isomorphism.

Let $S \rightarrowtail A$ be a multiplicative subobject, i.e., S satisfies $e \in S$, and $s, t \in S$ implies $s \cdot t \in S$. This means that we have factorizations

$$\begin{array}{ccc} & & S \\ & \nearrow & \downarrow \\ 1 & \xrightarrow{e} & A \end{array} \qquad \text{and} \qquad \begin{array}{ccc} S \times S & \dashrightarrow & S \\ \downarrow & & \downarrow \\ A \times A & \xrightarrow{\ \cdot\ } & A \end{array}$$

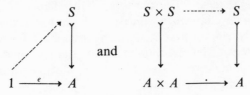

Equivalently, in terms of the characteristic map $\varphi: A \to \Omega$ of S, we have: $\varphi(e) = \text{true}$ and $\varphi(s) \wedge \varphi(t) \leq \varphi(s \cdot t)$.

We want to localize at S. That is, we want, in $A \times S$, to divide out by the equivalence relation: $(a, s) \sim (a', s')$ iff $\exists t \in S\ (tas' = ta's)$. Thus, we first interpret the latter formula in $(A \times S) \times (A \times S)$ by means of the diagram

$$\begin{array}{ccccccc} E' & \overset{i}{\rightarrowtail} & S \times (A \times S) \times (A \times S) & \underset{S \times \pi_{23}}{\overset{S \times \pi_{14}}{\rightrightarrows}} & S \times (A \times S) & \overset{\cdot}{\longrightarrow} & A \\ \downarrow & & \downarrow{\scriptstyle \pi} & & & & \\ E & \overset{\langle \rho_1, \rho_2 \rangle}{\rightarrowtail} & (A \times S) \times (A \times S) & & & & \end{array}$$

where i is an equalizer, and $\langle \rho_1, \rho_2 \rangle$ is its image under the projection π. As usual, we verify that E *is* an equivalence relation, and we form the coequalizer

$$E \underset{\rho_2}{\overset{\rho_1}{\rightrightarrows}} A \times S \overset{q}{\longrightarrow} A_S .$$

A_S, which inherits its ring structure from A, is A *localized at* S. Let e_A be the composite $A \to 1 \xrightarrow{e} A$, and ρ the composite $\langle id_A, e_A \rangle: A \to A \times S \xrightarrow{q} A_S$. Then ρ is a homorphism, and is universal among homomorphisms that take S to units. That is, we have

$$\begin{array}{ccc} S & \longrightarrow & U(A_S) \\ \downarrow & & \downarrow \\ A & \overset{\rho}{\longrightarrow} & A_S \end{array}$$

and for any other homorphism $\varphi: A \to A'$ such that

$$\begin{array}{ccc} S & \longrightarrow & U(A') \\ \downarrow & & \downarrow \\ A & \overset{\varphi}{\longrightarrow} & A' \end{array}$$

there is a unique homomorphism $\varphi': A_S \to A'$ such that

$$\begin{array}{ccc} A & \overset{\rho}{\longrightarrow} & A_S \\ & {\scriptstyle \varphi}\searrow \quad \swarrow{\scriptstyle \varphi'} & \\ & A' & \end{array}$$

commutes. These results are established essentially as in the classical case.

An important property of localization is given by

1.1. Proposition.

$$\begin{CD} S @>>> U(A_S) \\ @VVV @VVV \\ A @>>{\rho}> A_S \end{CD}$$

is a pullback iff S is *saturated*, i.e., satisfies $s, t \in S$ iff $st \in S$ or,

$$\begin{CD} S \times S @>>> S \\ @VVV @VVV \\ A \times A @>>{\cdot}> A \end{CD}$$

is a pullback.

Proof. Since $U(A_S)$ is clearly saturated, the necessity is obvious. If S is saturated, we prove

$$\begin{CD} S \times S @>>> U(A_S) \\ @VVV @VVV \\ A \times S @>{q}>> A_S \end{CD}$$

is a pullback, which is clearly sufficient. Leaving the commutativity to the reader, the classical proof goes as follows: If $a/s \in U(A_S)$ then there exists a'/s' such that $(a'/s') \cdot (a/s) = a'a/s's = e/e$. Thus, there exists $t \in S$ such that $t(a'a - s's) = 0$. Hence $t(a'a) \in S$, so $a \in S$. For purposes of illustration, we show how to translate this proof to an arbitrary topos. Thus, suppose we have a map $\langle a, s\rangle \colon X \to A \times S$ such that

$$\begin{CD} X @>>> U(A_S) \\ @V{\langle a, s\rangle}VV @VVV \\ A \times S @>{q}>> A_S \end{CD}$$

That is, we have $\alpha \colon X \to A_S$ such that $\alpha(q\langle a, s\rangle) = e_X$. Let

$$\begin{CD} X' @>{\langle a', s'\rangle}>> A \times S \\ @V{q'}VV @VV{q}V \\ X @>{\alpha}>> A_S \end{CD}$$

be a pullback, and call $X' \xrightarrow{q'} X \xrightarrow{\langle a, s\rangle} A \times S$ again $\langle a, s\rangle$ (restricted to X'). By definition of the multiplication on A_S, we have that the two composites

$$\begin{array}{ccccc} X' & \xrightarrow{\langle a'a,\, s's\rangle} & A \times S & \xrightarrow{q} & A_S \\ & \searrow & \nearrow {\scriptstyle \langle e, e\rangle} & & \\ & & 1 & & \end{array}$$

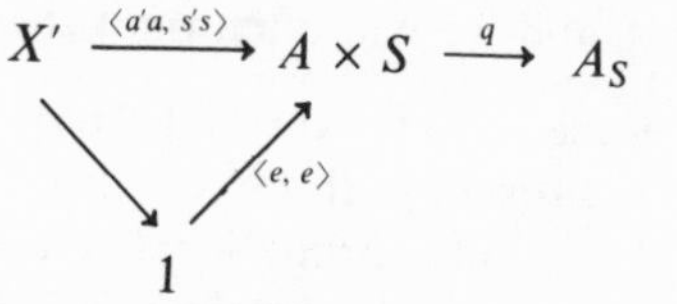

are equal. Since E is an equivalence relation, we have

$$\begin{array}{ccc} E & \xrightarrow{\langle \rho_1, \rho_2\rangle} & (A \times S) \times (A \times S) \\ & \nwarrow & \uparrow {\scriptstyle (\langle a'a,\, s's\rangle,\, \langle e, e\rangle)} \\ & & X' \end{array}$$

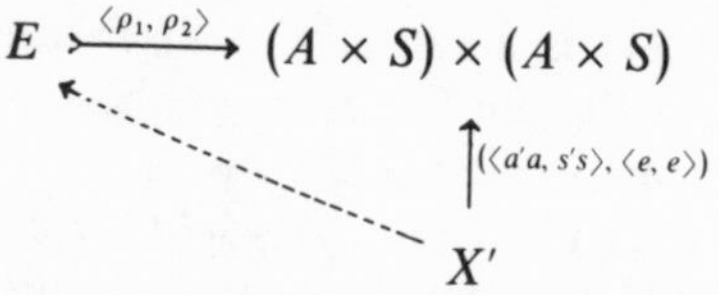

Take pullbacks again, using the same convention for restriction. We obtain

$$\begin{array}{ccccc} X'' & \xrightarrow{(t,\, \langle a'a,\, s's\rangle,\, \langle e, e\rangle)} & E' & \rightarrowtail & S \times (A \times S) \times (A \times S) \\ {\scriptstyle q''}\downarrow & & \downarrow & & \downarrow {\scriptstyle \pi} \\ X' & \xrightarrow[(\langle a'a,\, s's\rangle,\, \langle e, e\rangle)]{} & E & \rightarrowtail & (A \times S) \times (A \times S) \end{array}$$

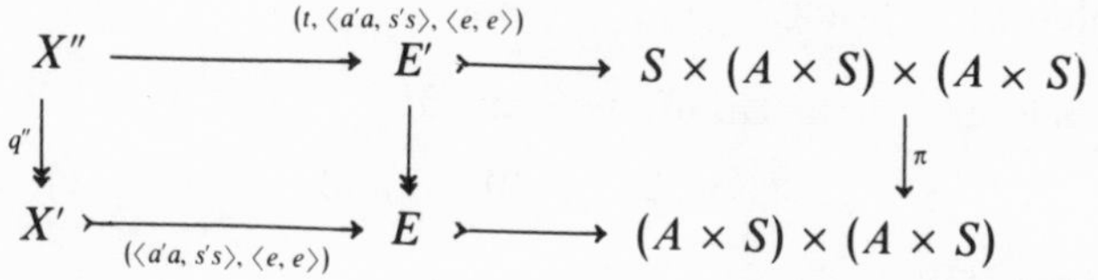

such that $t(a'a) = t(s's)$. Thus, as before, a (restricted to X'') is in S, i.e., we have

$$\begin{array}{ccccccc} & & & & & & S \\ & & & & & \nearrow & \downarrow \\ X'' & \underset{q''}{\twoheadrightarrow} & X' & \underset{q'}{\twoheadrightarrow} & X & \xrightarrow[a]{} & A \end{array}$$

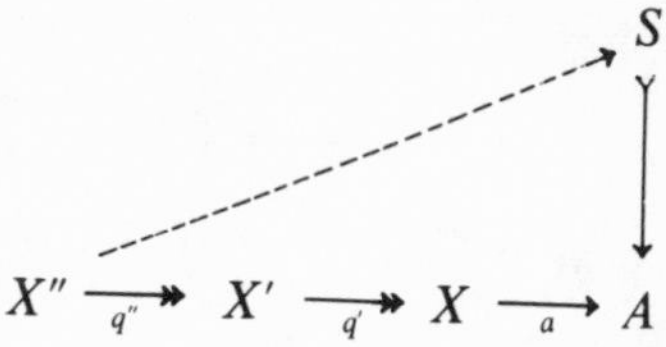

But then

$$\begin{array}{ccc} & & S \\ & \nearrow & \downarrow \\ X & \xrightarrow[a]{} & A \end{array}$$

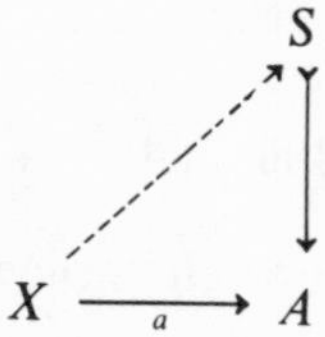

so that we have

$$\begin{array}{ccccc} & & S \times S & \longrightarrow & U(A_S) \\ & \nearrow & \downarrow & & \downarrow \\ X & \xrightarrow[\langle a, s\rangle]{} & A \times S & \underset{q}{\twoheadrightarrow} & A_S \end{array}$$

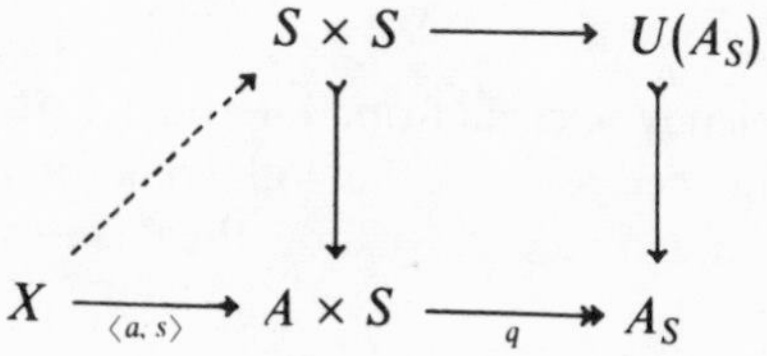

which finishes the proof. Notice that there are precisely as many restrictions to "covers" as there are instances of the existential quantifier in the classical proof. These translations can be formalized by means of Kripke–Joyal semantics, and we will often omit them.

We shall define a *prime* of A (as distinct from a prime ideal of A) to be a subobject $P \rightarrowtail A$ satisfying those conditions which, in the case $\mathbf{E} = \mathbf{S}$, define the *complement* of a prime ideal. We do this because it is at these objects that we wish to localize, and since $\neg\neg \neq id$, we must deal with them directly. Thus,

1.2. Definition. $P \rightarrowtail A$ is a *prime of* A iff P satisfies: $e \in P$, $fg \in P$ iff $f \in P$ and $g \in P$, not $(0 \in P)$, and $f + g \in P$ implies $f \in P$ or $g \in P$. In terms of the characteristic map $\varphi\colon A \to \Omega$ this says: $\varphi(e) = \text{true}$, $\varphi(fg) = \varphi(f) \wedge \varphi(g)$, $\varphi(0) = \text{false}$, and $\varphi(f + g) \le \varphi(f) \vee \varphi(g)$. The statement in terms of diagrams should be clear. Note that the description in terms of the characteristic map shows that the pullback of a prime along a ring homomorphism is a prime.

1.3. Definition. A is *local* iff the formulas

$$\forall f(f \in U(A) \vee (e - f) \in U(A)) \qquad \text{and} \qquad \neg(0 = e)$$

are valid. That is, if

$$\begin{array}{ccc} \overline{U}(A) & \longrightarrow & U(A) \\ \downarrow & & \downarrow \\ A & \xrightarrow[e_A - id_A]{} & A \end{array}$$

is a pullback, then A is local iff $\overline{U}(A) \vee U(A) \rightarrowtail A$ is an isomorphism, and $\varnothing \rightarrowtail 1 \underset{e}{\overset{0}{\rightrightarrows}} A$ is an equalizer.

The following is easily established.

1.4. Proposition. A is local iff $U(A) \rightarrowtail A$ is a prime.

The fundamental relation between "prime" and "local" is given by

1.5. Proposition. If $S \rightarrowtail A$ is a saturated multiplicative subobject, then S is prime iff A_S is local.

Proof. The sufficiency is clear from 1.4 and 1.1. If $S \rightarrowtail A$ is prime, the classical proof for the necessity is the following: if $a/s \in A_S$, then a or $s - a \in S$. If $a \in S$ we are done, and if not, then $(s - a)/s$ is a unit in A_S, but $e - a/s = (s - a)/s$. At the risk of being pedantic, we once again carry this proof over to an arbitrary topos for the benefit of those readers (should there be any!) who are unaccustomed to this sort of mathematics.

Thus, let $\tau\colon A \times S \to S \times A$ denote the switching map, and $-\colon A \times A \to A$ the subtraction on A. Put $\sigma = - \circ \tau$, and form the pullback

$$\begin{array}{ccc}
\bar{S} & \longrightarrow & S \\
\downarrow & & \downarrow \\
A \times S & \xrightarrow{\ \sigma\ } & A
\end{array}$$

$$\bar{S} = \text{“}\{\langle a, s\rangle \mid s - a \in S\}\text{”}.$$

Since

$$\begin{array}{ccc}
A \times S & \xrightarrow{\ q\ } & A_S \\
{\scriptstyle \langle \sigma, \pi_2 \rangle}\downarrow & & \downarrow{\scriptstyle e_{A_S} - id_{A_S}} \\
A \times S & \xrightarrow[\ q\ ]{} & A_S
\end{array}$$

commutes, it follows from 1.1 that

$$\begin{array}{ccc}
\bar{S} & \longrightarrow & \bar{U}(A_S) \\
\downarrow & & \downarrow \\
A \times S & \xrightarrow[\ q\ ]{} & A_S
\end{array}$$

is a pullback. We prove that $\bar{S} \vee (S \times S) \underset{\cong}{\rightarrowtail} A \times S$, which is sufficient by the preceding and 1.1. Thus, consider the diagram

$$\begin{array}{ccccc}
 & & S & & \\
 & & \downarrow & & \\
A \times S & \overset{\sigma}{\underset{\pi_1}{\rightrightarrows}} & A & \xrightarrow{\ \varphi\ } & \Omega
\end{array}$$

where φ is the characteristic map of $S \rightarrowtail A$. Now $\sigma + \pi_1 = \pi_2$, hence $\varphi(\sigma + \pi_1) = \text{true}$, so that $\varphi(\sigma) \vee \varphi(\pi_1) = \text{true}$, which is the result we want. Since $\varphi(0) = 0$, the validity of $\neg(0 = e)$ in A_S is clear. The most important property of localization, as I hope to indicate in a later paper, is now just the following remark.

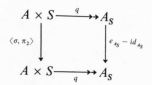

1.6. Proposition. If $\varphi: A \to A'$ is a homomorphism of A into a local ring A', then there is a universal factorization

$$\begin{array}{ccc} A & \xrightarrow{\varphi} & A' \\ & \searrow \quad \nearrow_{\varphi'} & \\ & \overline{A} & \end{array}$$

where $\overline{A}$ is local, and φ' is a local homomorphism.

φ' *local* means that

$$\begin{array}{ccc} U(\overline{A}) & \longrightarrow & U(A') \\ \downarrow & & \downarrow \\ \overline{A} & \xrightarrow[\varphi']{} & A' \end{array}$$

is a pullback, and *universal* means that for any other such factorization

$$\begin{array}{ccc} A & \xrightarrow{\varphi} & A' \\ & \searrow \quad \nearrow_{\varphi''} & \\ & \overline{\overline{A}} & \end{array}$$

there is a unique local homorphism $\overline{A} \to \overline{\overline{A}}$ making the resulting two triangles commute.

Proof. Let

$$\begin{array}{ccc} S & \longrightarrow & U(A') \\ \downarrow & & \downarrow \\ A & \xrightarrow{\varphi} & A' \end{array}$$

be a pullback. S is prime, so by 1.5 A_S is local. Taking $\overline{A} = A_S$, we have the obvious φ'. An easy exercise shows φ' is local. The universality of the factorization is a direct consequence of the universal property of $A \to \overline{A}$.

2. The Problem of the Spectrum and Some Previous Solutions

Let A be a commutative ring in **E**. We can look at the problem of finding the spectrum of A as the problem of finding a free local ring on A. That is, we try to find a universal homomorphism $\gamma: A \to L(A)$, where $L(A)$ is a local ring. Here, universal means that for any other such homomorphism $A \xrightarrow{\varphi} L'$, there exists a unique local homomorphism $L(A) \xrightarrow{\bar{\varphi}} L'$ such that

$$\begin{array}{ccc} A & \xrightarrow{\gamma} & L(A) \\ & {}_{\varphi}\searrow \quad \swarrow_{\bar{\varphi}} & \\ & L' & \end{array}$$

commutes. Now if the theory of local rings were an equational theory—and if $\mathbf{E}$ had a natural number object, which we will also need later—then we could find such an $L(A)$ (Lesaffre [7]). That is, at least for finitely presented algebraic theories, free algebras exist. Clearly, however, the theory of local rings is *not* equational, and just as clearly, the problem as stated cannot be solved, at least if we remain in the fixed topos $\mathbf{E}$. Admitting, then, the possibility of changing the topos $\mathbf{E}$, we need morphisms of ringed topoi. Thus, if $(\mathbf{E}', A')$ and $(\mathbf{E}, A)$ are ringed topoi, we define a *morphism* $(\mathbf{E}', A') \to (\mathbf{E}, A)$ to be a pair (f, φ), where $f: \mathbf{E}' \to \mathbf{E}$ is a geometric morphism, and $\varphi: f^*A \to A'$ is a homomorphism of rings. Note the duality involved here, which is standard in algebraic geometry. If $(\mathbf{E}', A')$ and $(\mathbf{E}, A)$ are local ringed topoi, meaning A' and A are local rings, we call (f, φ) *local* iff φ is local. Defining composition in the obvious way, we obtain two categories *R-top* and *LR-top*. Now, taking account of duality, let us ask the previous question in this broader context. Thus, given $(\mathbf{E}, A)$, we want to find a universal morphism of ringed topoi $(L(\mathbf{E}, A), LA) \xrightarrow{(\Gamma, \gamma)} (\mathbf{E}, A)$ where LA is local. Again, universal means that for any other morphism $(\mathbf{E}', L') \xrightarrow{(f, \varphi)} (\mathbf{E}, A)$ with L' local, there is a unique *local morphism* $(\bar{f}, \bar{\varphi}): (\mathbf{E}', L') \to (L(\mathbf{E}, A), LA)$ such that $(\Gamma, \gamma) \circ (\bar{f}, \bar{\varphi}) = (f, \varphi)$. That is, we are looking for a right adjoint to the forgetful functor LR-top $\to R$-top. When such an adjoint exists, it is characterized up to unique local equivalence, and we call it Spec$(\mathbf{E}, A)$, the *spectrum* of $(\mathbf{E}, A)$.

Before proceeding further, I should mention a technical result, known as Diaconescu's theorem, which will be needed in the sequel. Thus, let $\mathbf{E}$ be a topos and $\mathbf{C} \in \mathrm{Cat}(\mathbf{E})$ a category in $\mathbf{E}$. $\mathbf{C}$ consists of an object of objects C_0, an object of morphisms C_1, domain and codomain maps $\langle d_0, d_1 \rangle: C_1 \to C_0 \times C_0$, and a composition with identities. A contravariant, internal functor from $\mathbf{C}$ to $\mathbf{E}$ is an object $F \to C_0$ together with an operation $\xi: C_1 \times_{d_1} F \to F$ such that

$$\begin{array}{ccccc} F & \xleftarrow{\xi} & C_1 \times_{d_1} F & \xrightarrow{\pi_2} & F \\ \downarrow & & \downarrow{\scriptstyle \pi_1} & Pb & \downarrow \\ C_0 & \xleftarrow[d_0]{} & C_1 & \xrightarrow[d_1]{} & C_0 \end{array} \qquad (1)$$

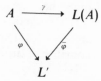

commutes, and the evident equations involving ξ with composition and identities are satisfied. When $\mathbf{E} = \mathbf{S}$, these are ordinary small categories and set-valued functors. Forming the category of functors and natural transformations yields a topos $\mathbf{E}(\mathbf{C}^{\text{op}})$ together with a geometric morphism $\Gamma: \mathbf{E}(\mathbf{C}^{\text{op}}) \to \mathbf{E}$; details can be found in [8]. Diaconescu's theorem [2] states then that if $f: \mathbf{E}_0 \to \mathbf{E}$ is a geometric morphism, liftings $\bar{f}: \mathbf{E}_0 \to \mathbf{E}(\mathbf{C}^{\text{op}})$ such that

$$\begin{array}{ccc} \mathbf{E}_0 & \xrightarrow{\bar{f}} & \mathbf{E}(\mathbf{C}^{\text{op}}) \\ & {\scriptstyle f}\searrow \quad \swarrow{\scriptstyle \Gamma} & \\ & \mathbf{E} & \end{array}$$

commutes, are in 1–1 correspondence with flat, covariant, internal functors from $f^*\mathbf{C}$ to $\mathbf{E}_0$. Here a functor is *flat* iff the category given by the top line in (1) is *cofiltered*, which definition must be taken in the strictly internal, but obvious, sense. We get one half of the correspondence by observing that there is a generic, covariant, flat functor $Y \to \Gamma^*(\mathbf{C})$. Namely, $\Gamma^*(\mathbf{C})$ in $\mathbf{E}(\mathbf{C}^{\text{op}})$ has as object of objects $\pi_2: C_0 \times C_0 \to C_0$, and the object assignment $Y \to \Gamma^*(\mathbf{C})_0$ is simply

$$\begin{array}{ccc} C_1 & \xrightarrow{\langle d_1, d_0 \rangle} & C_0 \times C_0 \\ & {\scriptstyle d_0}\searrow \quad \swarrow{\scriptstyle \pi_2} & \\ & C_0 & \end{array}$$

Given $\bar{f}$, we obtain $F \to f^*\mathbf{C}$ by evaluating $\bar{f}^*$ on $Y \to \Gamma^*(\mathbf{C})$. This works because inverse images preserve flatness. On the other hand, if $F \to f^*\mathbf{C}$ is an internal, flat functor from $f^*\mathbf{C}$ to $\mathbf{E}_0$, there is a unique $\bar{f}$—whose inverse image is "tensoring with F"—such that $\bar{f}^*(Y \to \Gamma^*(\mathbf{C}))$ is $F \to f^*\mathbf{C}$. An important consequence of this is that the following diagram is a pullback in the category of elementary topoi and geometric morphisms:

$$\begin{array}{ccc} \mathbf{E}_0(f^*\mathbf{C}^{\text{op}}) & \longrightarrow & \mathbf{E}(\mathbf{C}^{\text{op}}) \\ {\scriptstyle \Gamma_0}\downarrow & & \downarrow{\scriptstyle \Gamma} \\ \mathbf{E}_0 & \xrightarrow[f]{} & \mathbf{E} \end{array}$$

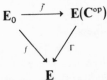

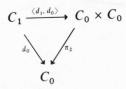

Let us now consider several previous approaches to the construction of the spectrum of a ringed topos $(\mathbf{E}, A)$. First that of [5]. Here one begins by internalizing the definition of "prime," say in terms of characteristic maps, to construct the object of primes $X \rightarrowtail \Omega^A$. It is easy to see that for $U \in \mathbf{E}$, the sections of X over U, i.e., the maps $U \to \Omega^A$ that factor through X, correspond precisely, by exponential adjointness, to the primes in the ring $U \times A \to U$ of $\mathbf{E}/U$. In particular, $X \rightarrowtail \Omega^A$ gives a canonical prime $P \rightarrowtail X \times A$ in $\mathbf{E}/X$ with characteristic map $\varphi: X \times A \to \Omega$ the transpose of $X \rightarrowtail \Omega^A$. Reversing φ and taking the transpose gives $\bar{\varphi}: A \to \Omega^X$. When $\mathbf{E} = \mathbf{S}$, $\bar{\varphi}(f) = D(f) = \{P \mid f \in P\}$; remember that these primes P are the *complements* of prime ideals. By pullback, $\bar{\varphi}$ induces a preorder $\langle d_0, d_1 \rangle: S \rightarrowtail A \times A$ on A: $(f, g) \in S$, or $f \leq g$, iff $D(f) \leq D(g)$. Let $\mathbf{A}$ in $\mathrm{Cat}(\mathbf{E})$ be A with the above category structure. In $\mathbf{E}(\mathbf{A}^{\mathrm{op}})$ the generic flat functor is

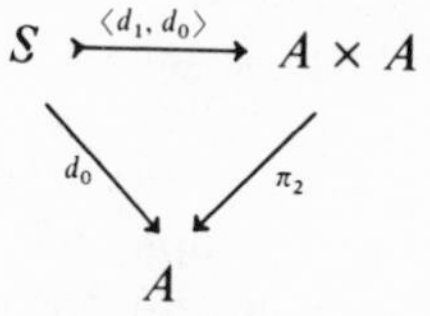

and in $\mathbf{E}/X$ we have the canonical prime

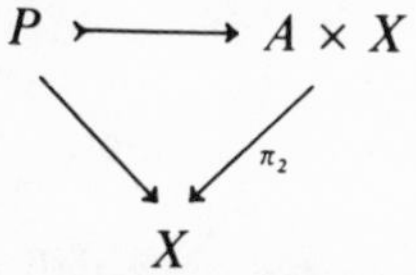

which is a flat functor. Thus we obtain a unique geometric morphism:

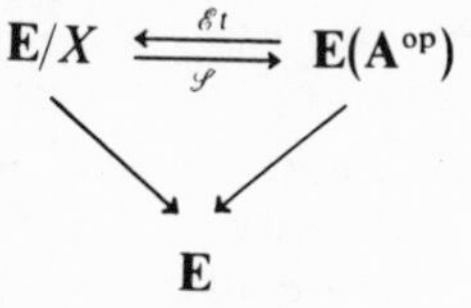

such that $\mathscr{E}t(S \rightarrowtail \Gamma^* A) = P \rightarrowtail A \times X$. Factoring this morphism gives the spectrum:

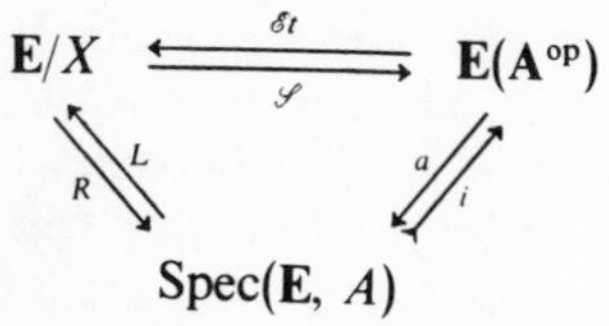

Note that the objects of $\mathrm{Spec}(\mathbf{E}, A)$ can be thought of as objects with structure over X—i.e., “éspaces étalés” over X—or as sheaves on the basis consisting of the $D(f)$. The local ring in Spec(**E**, A) is obtained as follows: $\mathscr{E}t(S \rightarrowtail \Gamma^* A) = L(aS \rightarrowtail a\Gamma^* A) = P \rightarrowtail A \times X$ is a prime. But L is faithful, and hence reflects the notion of prime. Thus, $aS \rightarrowtail a\Gamma^* A$ is a prime, and the required local ring is $(a\Gamma^* A)_{aS}$.

Because the necessary technical means—e.g., Diaconescu's theorem—were not then available, the universal property of this simple construction, obviously correct in the case $\mathbf{E} = \mathbf{S}$, was difficult to establish. As a result, a complete proof was never given, and it remained for Joyal to point out that, in fact, this construction is incorrect in the general case.

Joyal proceeded as follows. As in [3], let **R** be the category of commutative rings of finite type. $\mathbf{R}^{\text{op}}$ has finite limits, so that in any Grothendieck topos $\mathbf{E} \xrightarrow{\Gamma} \mathbf{S}$, the flat **E**-valued functors on $\Gamma^*\mathbf{R}^{\text{op}}$ are the (internal) finite limit preserving **E**-valued functors on $\Gamma^*\mathbf{R}^{\text{op}}$. These correspond 1–1 to the *external* finite limit preserving functors $F\colon \mathbf{R}^{\text{op}} \to \mathbf{E}$. Such an F, however, is completely determined by the ring $F(\mathbb{Z}[t])$ in **E**. In particular, in the universal example $\mathbf{S}^{\mathbf{R}}$, the universal flat functor corresponds to the representable functor $\mathbf{R}(\mathbb{Z}[t],\ )$, which is the underlying functor $A\colon \mathbf{R} \to \mathbf{S}$. The ring structure on the latter is given object by object in the obvious manner. Thus, $(\mathbf{S}^{\mathbf{R}}, A)$ is the universal ringed topos for topoi defined over **S**. What Joyal did was to calculate the object of primes $X \rightarrowtail \Omega^A$ in $\mathbf{S}^{\mathbf{R}}$, and show that $X = \varnothing$, which devastates the previous construction. Let us consider, for a moment, the details of calculations such as these.

In any ringed topos (**E**, A), we have defined a *prime* $P \rightarrowtail A$ to be a subobject whose characteristic map $\varphi\colon A \to \Omega$ satisfies $\varphi(e) = \text{true}$, $\varphi(fg) = \varphi(f) \wedge \varphi(g)$, $\varphi(0) = \text{false}$, and $\varphi(f+g) \leq \varphi(f) \vee \varphi(g)$. Let us weaken these and consider $I \rightarrowtail A$ whose characteristic map $\varphi\colon A \to \Omega$ satisfies $\varphi(0) = \text{false}$, $\varphi(f+g) \leq \varphi(f) \vee \varphi(g)$, and $\varphi(fg) \leq \varphi(f) \wedge \varphi(g)$—these are the properties that characterize the *complement of an ideal* in case $\mathbf{E} = \mathbf{S}$. Or we may consider $\mathfrak{P} \rightarrowtail A$ such that $\varphi(0) = \text{true}$, $\varphi(e) = \text{false}$, $\varphi(fg) = \varphi(f) \vee \varphi(g)$, and $\varphi(f) \wedge \varphi(g) \leq \varphi(f+g)$. These, of course, are the axioms for a *prime ideal* of A. As with the object of primes $X \rightarrowtail \Omega^A$, we can construct the object of complements of ideals $Y \rightarrowtail \Omega^A$, and the object of prime ideals $Z \rightarrowtail \Omega^A$.

We prove now that in the universal example $(\mathbf{S}^{\mathbf{R}}, A)$, X and Z are empty, and Y is $\{\varnothing\}\colon 1 \to \Omega^A$. There is, first of all, a trivial reason for this, and that is that $\neg(0 = e)$ is invalid. In fact, $[\![0 = e]\!]$ is the subobject of 1 that takes the value 1 at the zero ring (the terminal object where $1 = 0$), and $\varnothing$ everywhere else. Clearly, there can be no such dichotomies as in the definition of prime or prime ideal unless $[\![0 = e]\!] = \varnothing$. Similarly, the only complement of an ideal is $\varnothing$. But this is a trivial reason, since one can either force the predicate $\neg(0 = e)$—as I will indicate later—or else we can change the definitions in the obvious way. The real trouble is naturality. Namely, to have one of these objects nonempty we must have a section whose domain is a representable functor R. This corresponds to a structure of the given type in the ring $R \times A \to R$ in $\mathbf{S}^{\mathbf{R}}/R$. But $\mathbf{S}^{\mathbf{R}}/R$ is equivalent to $\mathbf{S}^{R/\mathbf{R}}$, and under this equivalence $R \times A \to R$ becomes the ring A_R: $R/\mathbf{R} \to \mathbf{S}$ given by $A_R(R \xrightarrow{\varphi} S) = A(S) = S$. Now each of the three sets of conditions are object by object, so to have the complement of an ideal $I \rightarrowtail A_R$, for example, means to have a natural complement $I(S)$ in each R-algebra S of finite type. Suppose $S \neq 1$ (the terminal ring with $1 = 0$), and $I(S) \neq \varnothing$. Then $U(S) \rightarrowtail I(S)$. Suppose $x \in I(S)$ is a nonunit. Then $x \in \mathfrak{M}$ for $\mathfrak{M}$ some maximal ideal. Let $k = S/\mathfrak{M}$. Then we must have

$$\begin{array}{ccccc} R & \xrightarrow{\varphi} & S & \leftarrowtail & I(S) \\ & \searrow & \downarrow & & \downarrow \\ & & k & \leftarrowtail & I(k) \end{array}$$

by naturality. There are only two possibilities for $I(k)$: $I(k) = \varnothing$, or $I(k) = U(k)$—the nonzero elements of k. We cannot have $I(k) = \varnothing$, since $I(S) \neq \varnothing$. But also, we can not have $I(k) = U(k)$, since $x \in I(S)$ and goes to zero under $S \to k$. Thus, either $I(S) = \varnothing$, or $I(S) = U(S)$ and S is local. If S is local, then for every homomorphism $\psi: S \to S'$ we must have $I(S') \neq \varnothing$, i.e., S' is local. But if $\mathfrak{M}$ is the maximal ideal of S, and $k = S/\mathfrak{M}$, then we have the obvious homomorphism $S \to k[t]$, and $k[t]$ is *not* local. Thus, each $I(S) = \varnothing$ and Y is $\{\varnothing\}$. If the above I had been a prime, then it would have satisfied $e \in I$, so X is $\varnothing$. To give a prime ideal $\mathfrak{P} \rightarrowtail A_R$ would be to give in each R-algebra S of finite type a natural prime ideal $\mathfrak{P}(S) \rightarrowtail S$. If $S \neq 1$, let $\mathfrak{p}$ be a prime ideal of S. Suppose $x \in \mathfrak{P}(S)$, but $x \notin \mathfrak{p}$. Then we have

$$\begin{array}{ccccc} R & \xrightarrow{\varphi} & S & \leftarrowtail & \mathfrak{P}(S) \\ & \searrow & \downarrow & & \downarrow \\ & & S/\mathfrak{p} & \leftarrowtail & \mathfrak{P}(S/\mathfrak{p}) \end{array}$$

with an $\bar{x} \neq 0$ in $\mathfrak{P}(S/\mathfrak{p})$. Let S' be $S/\mathfrak{p}$ localized at the multiplicative subset determined by $\bar{x}$. Then, since $S/\mathfrak{p}$ is an integral domain, S' is not the zero ring and we have

$$\begin{array}{ccc} S/\mathfrak{p} & \leftarrowtail & \mathfrak{P}(S/\mathfrak{p}) \\ {\scriptstyle p}\downarrow & & \downarrow \\ S' & \leftarrowtail & \mathfrak{P}(S') \end{array}$$

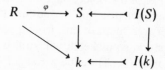

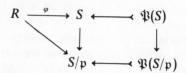

But under p, $\bar{x}$ goes into $\mathfrak{P}(S')$ and also becomes a unit, which is impossible. Thus, $\mathfrak{P}(S)$ is the intersection of all prime ideals of S, and hence the nilpotent elements of S form a prime ideal. Furthermore, this must be the case for every S-algebra of finite type. But this is absurd, as witness, say, $\Delta: S \to S \times S$. If $x = (e, 0)$ and $y = (0, e)$, then $xy = 0$, but neither x nor y is nilpotent. Thus, Z is also empty.

In view of the preceding calculations, it was necessary to find a new way to construct the spectrum of a ringed topos $(\mathbf{E}, A)$. Joyal suggested a method based on the construction of a certain universal distributive lattice object in $\mathbf{E}$ associated to A, but again, no details were given. (See [6], however, for a description of the proposed construction.) Later, when Joyal showed me his calculation of the object of primes in the universal example, I found another related construction based on the notion of "radical ideal." I will give a fuller treatment of this, from a somewhat different point of view, in a separate paper, so a sketch will suffice for now.

Thus, in a ringed topos $(\mathbf{E}, A)$ define a *radical ideal* to be a subobject $\mathfrak{A} \rightarrowtail A$ such that $0 \in \mathfrak{A}$, $f, g \in \mathfrak{A}$ implies $f + g \in \mathfrak{A}$, $f \in \mathfrak{A}$ and $a \in A$ implies $af \in \mathfrak{A}$, and $f \in \mathfrak{A}$ iff $f^2 \in \mathfrak{A}$. In terms of the characteristic map $\varphi: A \to \Omega$ of $\mathfrak{A} \rightarrowtail A$, this means: $\varphi(0) = \text{true}$, $\varphi(f) \wedge \varphi(g) \le \varphi(f + g)$, $\varphi(f) \le \varphi(af)$, $\varphi(f) = \varphi(f^2)$. As before, we can easily construct the object of radical ideals $R \rightarrowtail \Omega^A$. Here, R is a retract of Ω^A, for we can define a retraction $r: \Omega^A \to R$ by $r(X \rightarrowtail A) = \inf\{\mathfrak{A} \mid X \le \mathfrak{A} \in R\}$. In fact, R is a complete Heyting algebra in $\mathbf{E}$, though this does not result from r preserving finite products, which is false.

Let us denote the composite $A \xrightarrow{\{\}} \Omega^A \xrightarrow{r} R$ by ρ. At least if $\mathbf{E}$ has a natural number object N, we can verify that ρ satisfies the following: $\rho(e) = A$ – the unit of R, $\rho(fg) = \rho(f) \wedge \rho(g)$, $\rho(f + g) \le \rho(f) \vee \rho(g)$, and $\rho(0) = \text{Nil}(A)$ – the zero of R. Moreover, ρ induces a preorder on A by pullback. Let $\mathbf{A}$ be the associated category in $\mathbf{E}$. Taking account of the structure of Ω in $\mathbf{E}(\mathbf{A}^{\text{op}})$ as the object of filters of elements of A under this preorder, we see that r itself induces a topology in $\mathbf{E}(\mathbf{A}^{\text{op}})$—the Zariski topology, of which more later. Passing to the category of sheaves gives Spec$(\mathbf{E}, A)$. Let Ω_A denote the Ω of Spec$(\mathbf{E}, A)$. One way to find the required homomorphism $\Gamma^*A \to L(A)$ with $L(A)$ local would be to localize Γ^*A at a (universal) prime $P \rightarrowtail \Gamma^*A$. To have such a prime we would need a map $\varphi: \Gamma^*A \to \Omega_A$ satisfying $\varphi(e) = \text{true}$, $\varphi(fg) = \varphi(f) \wedge \varphi(g)$, $\varphi(f + g) \le \varphi(f) \vee \varphi(g)$, and $\varphi(0) = \text{false}$. But these correspond 1–1 to maps $A \to \Gamma_*\Omega_A$, satisfying the associated conditions. Since the topology was induced by r, $\Gamma_*\Omega_A \simeq R$, and we have a canonical such map, namely ρ. This gives the local ring.

I should remark that, contrary to the previous structures, there are plenty of radical ideals in the universal example. In fact, the classical definition of the radical of an ideal generated by a subset of a ring is functorial if the subset is, so r is given object by object.

Simple as it is, there seemed to me to be two drawbacks to this construction. One was that apparently N is needed to verify the condition $\rho(fg) = \rho(f) \wedge \rho(g)$—basically this amounts to establishing the validity of the classical description of the radical of an ideal generated by a subobject. Another was that the technical details are quite complicated—especially those involving the definition of the topology and the universal property of Spec(**E**, A). Thus, towards the goal of removing these final snags, we will completely reconsider the problem in the next section.

3. A New Method

The construction we consider in this section is accomplished in two essentially independent stages. First we admit the apparent necessity of a natural number object and use it to put the desired order directly on the ring of **E** in the strongest possible way. Next we discuss the general concept of "Zariski topologies," and apply the result to the particular case at hand.

To begin, let (**E**, A) be a ringed topos with natural number object N. By primitive recursion, define the map $A \times N \to A$, which in **S**, takes (f, n) to f^n. Let $S \rightarrowtail A \times A$ be the interpretation of the formula $\exists n \exists a (f^n = ag) \wedge (n \geq 1)$.

3.1. Proposition. *S is a preorder, which provides the associated internal category **A** with (internal) finite inverse limits.*

Proof. I shall give the proof in **S**, assuming that by now the reader believes in a more or less formal method of translation.

Reflexivity being obvious, suppose we have $f \leq g$ and $g \leq h$. Then $f^n = ag$ and $g^m = bh$, so $(f^n)^m = a^m g^m = (a^m b)h$, and $f \leq h$.

Concerning finite limits in **A**, I claim e is a terminal object, and fg is $f \wedge g$, which will suffice. Well, $f = fe$, so $f \leq e$; $fg = fg$, so $fg \leq f, g$. Furthermore, suppose $h \leq f, g$. Then $h^n = af$, and $h^m = bg$, so $h^{n+m} = h^n h^m = (ab)fg$ and $h \leq fg$.

An important point about this category structure on A is that if $f\colon \mathbf{E}_0 \to \mathbf{E}$ is a geometric morphism, then the category structure $f^*S \rightarrowtail f^*A \times f^*A$ on f^*A is the above preorder on the ring f^*A given by the natural number object $N_0 \simeq f^*N$ of $\mathbf{E}_0$. Thus, if we can identify the flat (internal) functors from **A** to **E** in terms of A, the same identification will hold for any of the rings f^*A.

But it is well-known that for an internal category with finite limits, the flat functors are precisely the internal finite-limit-preserving ones. Thus, if

$F \to A$ is a flat functor, for any $I \in \mathbf{E}$ and any pair $I \overset{f}{\underset{g}{\rightrightarrows}} A$, denoting by $F(f) \to I$ the pullback

$$\begin{array}{ccc} F(f) & \longrightarrow & F \\ \downarrow & & \downarrow \\ I & \xrightarrow{f} & A \end{array}$$

we must have $F(e_I) = id_I$ and $F(fg) = F(f) \times_I F(g)$. Each $f \leq e$ is a monomorphism, so each $F(f) \to I$ is monic, which is equivalent to $F \rightarrowtail A$. Then, however, if $\varphi\colon A \to \Omega$ is the characteristic map of F, the above conditions become $\varphi(e) = \text{true}$, and $\varphi(fg) = \varphi(f) \wedge \varphi(g)$. Thus, $F \rightarrowtail A$ is a saturated multiplicative subobject. On the other hand, suppose $S \rightarrowtail A$ is a saturated multiplicative subobject with characteristic map $\psi\colon A \to \Omega$. If $f \leq g$ then $f^n = ag$ and $\psi(f) = \psi(f^n) = \psi(ag) = \psi(a) \wedge \psi(g) \leq \psi(g)$, so S is an internal, finite-limit-preserving functor.

Summing up the above, by Diaconescu's theorem we have

3.2. Theorem. If $\mathbf{E}_0 \xrightarrow{f} \mathbf{E}$ is a geometric morphism, liftings $\bar{f}$ making

$$\begin{array}{ccc} \mathbf{E}_0 & \xrightarrow{\bar{f}} & \mathbf{E}(\mathbf{A}^{\mathrm{op}}) \\ & {\scriptstyle f}\searrow \quad \swarrow{\scriptstyle \Gamma} & \\ & \mathbf{E} & \end{array}$$

commute are in 1–1 correspondence with the saturated multiplicative subobjects of f^*A. The correspondence is as follows: given $\bar{f}$ evaluate $\bar{f}^*$ on the universal saturated multiplicative subobject $S \rightarrowtail \Gamma^*A$.

Notice that the construction of $\mathbf{E}(\mathbf{A}^{\mathrm{op}})$ is stable under change of base and that it has nothing to do with the addition on A—Theorem 3.2 is a result about abelian monoids.

We start the next stage of the construction with some very general remarks on topologies in an arbitrary topos $\mathbf{E}$—no rings involved here.

First of all, as is fairly well known by now (see, say, [2]), to every subobject $I \rightarrowtail \Omega$, there is a smallest $J \rightarrowtail \Omega$ such that

$$\begin{array}{ccc} I & \rightarrowtail & J \\ & \searrow \quad \swarrow & \\ & \Omega & \end{array}$$

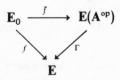

and the characteristic map $j: \Omega \to \Omega$ of J is a topology. In fact, J is the (internal) inf of all such subobjects of Ω; j is called the *topology generated by I*. If $X' \overset{m}{\rightarrowtail} X$ is an arbitrary monomorphism of $\mathbf{E}$, let $\varphi: X \to \Omega$ be its characteristic map, and denote by j_m the topology generated by the image I of φ. If $\mathbf{E}_0 \overset{f}{\to} \mathbf{E}$ is a geometric morphism, let $j_f: \Omega \to \Omega$ denote the topology defined by f—that topology whose sheaves form the image of f. Consider the diagram

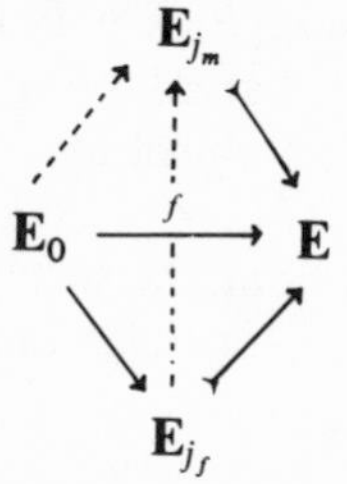

There is a factorization of f across $\mathbf{E}_{j_m} \rightarrowtail \mathbf{E}$ iff $\mathbf{E}_{j_f}$ is smaller than $\mathbf{E}_{j_m}$,

iff 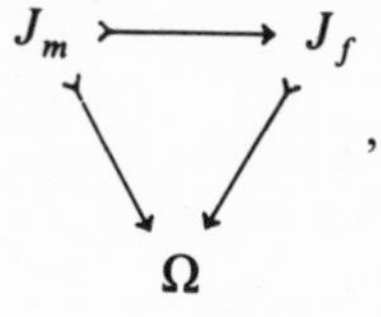 , iff 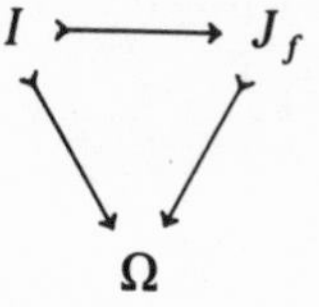

iff m is dense for j_f, iff $f^*X' \xrightarrow{f^*m} f^*X$ is an isomorphism. Note this shows immediately that

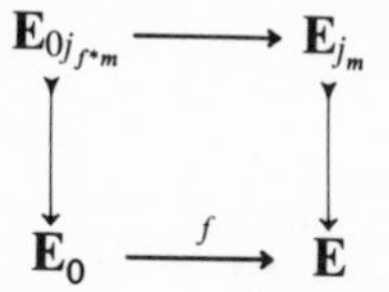

is a pullback—i.e., the construction of $\mathbf{E}_{j_m}$ is stable under change of base. Also, any topology $j: \Omega \to \Omega$ with corresponding $J \rightarrowtail \Omega$ arises in this way by taking the monomorphism m to be true: $1 \rightarrowtail J$.

Given any two monomorphisms $X' \overset{m}{\rightarrowtail} X$ and $Y' \overset{n}{\rightarrowtail} Y$, the appropriate topos is the intersection

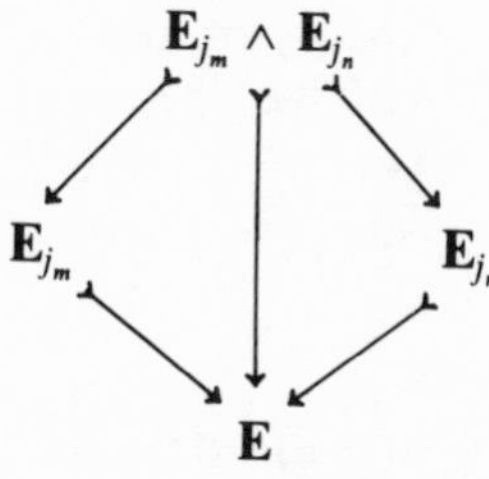

A geometric morphism factors across this iff its inverse image inverts both m and n.

How can one apply this? Well, suppose T is a finitely axiomatizable, coherent, or geometric theory in the sense of Reyes. That is, T has sorts $i, j, \ldots,$ function symbols f, and formulas φ built up from the atomic ones by means of the logical connectives: true, $\wedge$, $=$, $\vee$, $\exists$, false. The axioms of T consist of a finite number of sequents $\varphi \vdash \psi^V$, where φ and ψ are formulas of T and where V is a finite set of variables; rules of inference can be found in [1]. Let M be a structure for T in a topos $\mathbf{E}$—i.e., for each sort i we have an object M_i of $\mathbf{E}$, for each function symbol $f: i_1, \ldots, i_n \to i$ a map $M(f)$ from $M_{i_1} \times \cdots \times M_{i_n}$ to M_i, and for each atomic formula φ with free variables among the sorts $j_1, \ldots, j_k$ a subobject $M(\varphi)$ of $M_{j_1} \times \cdots \times M_{j_k}$. The rest of the formulas are interpreted in the standard manner. Now, if the axioms of T are $\varphi_1 \vdash \psi_1^{V_1}, \ldots, \varphi_m \vdash \psi_m^{V_m}$, let j_M be the least topology that forces $\varphi_1 \Rightarrow \psi_1, \ldots, \varphi_m \Rightarrow \psi_m$.

Perhaps a remark should be made here. Namely, suppose φ and ψ are two maps $X \to \Omega$ with corresponding subobjects $X' \rightarrowtail X$ and $X'' \rightarrowtail X$. Then $\varphi \Rightarrow \psi = \text{true}_X$ iff in

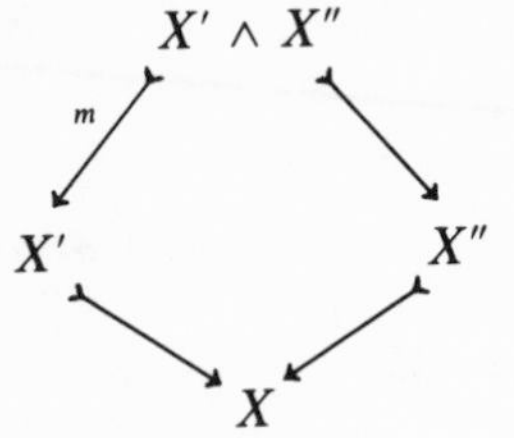

m is an isomorphism. Thus, the forcing topology for $\varphi \Rightarrow \psi$ is j_m. A particular case of this is well-known. In fact, suppose φ is a sentence $1 \to \Omega$ with truth value $U \rightarrowtail 1$. Then $\neg\varphi = \varphi \Rightarrow \text{false}$, and the forcing topology for $\neg\varphi$ is $U \vee (\ ): \Omega \to \Omega$, which gives the closed complement $\mathbf{E}_U \rightarrowtail \mathbf{E}$ as category of sheaves.

In any case, in the above situation $\mathbf{E}_{j_M} \rightarrowtail \mathbf{E}$ is the best topos in which M becomes a *model* of T—i.e., if $\mathbf{E}_0 \xrightarrow{f} \mathbf{E}$ is a geometric morphism, then f factors across $\mathbf{E}_{j_M} \rightarrowtail \mathbf{E}$ iff the structure f^*M is a model of T.

For example, let T be the theory of local rings. T has one sort i, the function symbols and axioms of the theory of commutative rings, and two further axioms:

$$\text{true} \vdash \exists a'(a'a = e) \vee \exists a''(a''(e - a) = e)^{\{a\}}$$

and

$$(0 = e) \vdash \text{false}.$$

Suppose $(\mathbf{E}, A)$ is a ringed topos. Then the interpretation of the formula $\exists a'(a'a = e) \vee \exists a''(a''(e - a) = e)$ is $\overline{U}(A) \vee U(A) \overset{m}{\rightarrowtail} A$, and the interpretation of $0 = e$ is the equalizer $V \rightarrowtail 1 \underset{0}{\overset{e}{\rightrightarrows}} A$. Call the best topology that inverts m and the canonical map $\varnothing \rightarrowtail V$ z_A—the *Zariski topology* corresponding to A. Then $\mathbf{E}_{z_A} \rightarrowtail \mathbf{E}$ has the following universal property: $f\colon \mathbf{E}_0 \to \mathbf{E}$ factors across $\mathbf{E}_{z_A} \rightarrowtail \mathbf{E}$ iff f^*A is local. Note that here, as in the general case above,

$$\begin{array}{ccc} \mathbf{E}_{0z_{f^*A}} & \longrightarrow & \mathbf{E}_{z_A} \\ \downarrow & & \downarrow \\ E_0 & \xrightarrow{f} & \mathbf{E} \end{array}$$

is a pullback.

Let us now use this to finish the construction of the spectrum. Thus, beginning with a ringed topos $(\mathbf{E}, A)$ with natural number object N, let $S \rightarrowtail A \times A$ be the previous preorder on A, and $\mathbf{A}$ the associated internal category of $\mathbf{E}$. As we have seen, in the topos $\mathbf{E}(\mathbf{A}^{\mathrm{op}}) \xrightarrow{\Gamma} \mathbf{E}$, $S \rightarrowtail \Gamma^*A$ is the universal saturated multiplicative subobject. We take the Zariski topology corresponding to $(\Gamma^*A)_S$ and call the resulting category of sheaves $\mathrm{Spec}(\mathbf{E}, A)$. Let us write LA for the sheaf associated to $(\Gamma^*A)_S$; LA is the local ring of $\mathrm{Spec}(\mathbf{E}, A)$. Call the morphism defining $\mathrm{Spec}(\mathbf{E}, A)$ over $\mathbf{E}$ again Γ. The preceding gives us the following universal property for the spectrum: If $f\colon \mathbf{E}_0 \to \mathbf{E}$ is a geometric morphism, then liftings $\bar{f}$ of f to $\mathbf{E}(\mathbf{A}^{\mathrm{op}})$ are in 1–1 correspondence with saturated multiplicative subobjects $\overline{S} \rightarrowtail f^*A$. Such a lifting $\bar{f}$ factors across $\mathrm{Spec}(\mathbf{E}, A)$ iff $\bar{f}^*(\Gamma^*A_S) \simeq (\bar{f}^*\Gamma^*A)_{\bar{f}^*S} \simeq f^*A_{\overline{S}}$ is local, iff $\overline{S} \rightarrowtail f^*A$ is prime. Thus, morphisms $\bar{f}$ such that

$$\begin{array}{ccc} \mathbf{E}_0 & \xrightarrow{\bar{f}} & \mathrm{Spec}(\mathbf{E}, A) \\ & {\scriptstyle f}\searrow \quad \swarrow{\scriptstyle \Gamma} & \\ & \mathbf{E} & \end{array}$$

commutes are in 1–1 correspondence with the primes of f^*A. In particular, the $\mathbf{E}$-points of $\mathrm{Spec}(\mathbf{E}, A)$—the $\bar{f}$ for $f = id_{\mathbf{E}}$—correspond 1–1 to the primes of A. Note that for any f,

$$\begin{array}{ccc} \mathrm{Spec}(\mathbf{E}_0, f^*A) & \longrightarrow & \mathrm{Spec}(\mathbf{E}, A) \\ {\scriptstyle \Gamma_0}\downarrow & & \uparrow{\scriptstyle \Gamma} \\ \mathbf{E}_0 & \xrightarrow[f]{} & \mathbf{E} \end{array}$$

is a pullback.

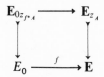

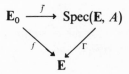

We denote by $\gamma: \Gamma^*A \to LA$ the canonical homomorphism. Thus, $(\mathrm{Spec}(\mathbf{E}, A), LA) \xrightarrow{(\Gamma, \gamma)} (\mathbf{E}, A)$ is a morphism of ringed topoi; let us show it has the correct universal property. So, suppose $(\mathbf{E}_0, L) \xrightarrow[(f, \varphi)]{} (\mathbf{E}, A)$ is a morphism of ringed topoi with L local. Consider the diagram

$$\begin{array}{ccc} \bar{P} & \longrightarrow & U(L) \\ \downarrow & & \downarrow \\ f^*A & \xrightarrow{\varphi} & L \\ & \searrow \quad \nearrow_{\bar{\varphi}} & \\ & (f^*A)_P & \end{array}$$

where the prime $\bar{P} \rightarrowtail f^*A$ is the pullback of $U(L) \rightarrowtail L$ by φ, and $\bar{\varphi}$ is local. Then $\bar{P}$ defines $\bar{f}$ such that

$$\begin{array}{ccc} \mathbf{E}_0 & \xrightarrow{\bar{f}} & \mathrm{Spec}(\mathbf{E}, A) \\ & {}_f\searrow \quad \swarrow_{\Gamma} & \\ & \mathbf{E} & \end{array}$$

commutes. The required morphism of local ringed topoi is thus $(\bar{f}, \bar{\varphi}): (\mathbf{E}_0, L) \to (\mathrm{Spec}(\mathbf{E}, A), LA)$.

To show uniqueness, let (f_0, φ_0) be another morphism of local ringed topoi such that

$$\begin{array}{ccc} (\mathbf{E}_0, L) & \xrightarrow{(f_0, \varphi_0)} & (\mathrm{Spec}(\mathbf{E}, A), LA) \\ & {}_{(f, \varphi)}\searrow \quad \swarrow_{(\Gamma, \gamma)} & \\ & (\mathbf{E}, A) & \end{array}$$

commutes. We write $P \rightarrowtail \Gamma^*A$ for the reflection of the universal multiplicative subobject S. Then P is prime and LA is $(\Gamma^*A)_P$. Since the diagram commutes above,

$$\begin{array}{ccc} f^*A & \xrightarrow{\varphi} & L \\ \wr\!\!\mid & & \uparrow \varphi_0 \\ f_0{}^*\Gamma^*A & \xrightarrow[f_0{}^*\gamma]{} & f_0{}^*LA \end{array}$$

commutes in $\mathbf{E}_0$. Because φ_0 is local, $f_0{}^*P \rightarrowtail f_0{}^*\Gamma^*A$ is the pullback of $U(L) \rightarrowtail L$ under φ—i.e., it is the previous $\bar{P} \rightarrowtail f^*A$. By Diaconescu's theorem then, $f_0 = \bar{f}$. $\varphi_0 = \bar{\varphi}$ by the uniqueness in the universal property of localization.

For completeness, two further remarks should be added to this treatment of the spectrum. One is that in **E** the canonical map $A \to \Gamma_*(LA)$ is an isomorphism—i.e., the representation of A in the ring of "global sections" of LA is complete. The second, due to Mulvey in the case $\mathbf{E} = \mathbf{S}$, is that in Spec(**E**, A) the formula

$$\neg(x \in U(LA)) \Rightarrow \exists n(x^n = 0)$$

is valid. This is surely important, though its precise significance is still somewhat obscure—as is the case with many such nongeometric formulas. In any case, calculations such as these are easier from the point of view of the Heyting algebra of radical ideals of A, and hence will be omitted here.

In conclusion, I would like to indicate briefly how this method of "forcing topologies" can be applied to other problems in topos theory. For example, take the problem of constructing classifying topoi. If T is a finitely presented algebraic theory, and **E** is an arbitrary base topos, Johnstone [4] constructs an internal category **T** in **E** such that **E**(**T**) is a T-algebra classifier for topos defined over **E**. More precisely, if $f: \mathbf{E}_0 \to \mathbf{E}$ is a geometric morphism, then T-algebras in $\mathbf{E}_0$ are in 1–1 correspondence with morphisms $\bar{f}$ such that

$$\begin{array}{ccc} \mathbf{E}_0 & \xrightarrow{\ \bar{f}\ } & \mathbf{E}(\mathbf{T}) \\ & {\scriptstyle f}\searrow \quad \swarrow{\scriptstyle \Gamma} & \\ & \mathbf{E} & \end{array}$$

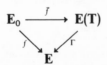

commutes. The correspondence is obtained by evaluating $\bar{f}^*$ on the universal T-algebra in **E**(**T**). When T is the theory of commutative rings, he then spends Chapter VI constructing the classifying topos for the theory of local rings by laboriously internalizing the construction of Hakim [3]. The correct solution, however, is extremely simple. Namely, if R is the universal ring, let z_R be the Zariski topology associated to R. Obviously, $\mathbf{E}(\mathbf{T})_{z_R}$ is the classifying topos for the theory of local rings: Commutative rings A in $\mathbf{E}_0$ correspond 1–1 to **E**-morphisms $\bar{f}: \mathbf{E}_0 \to \mathbf{E}(\mathbf{T})$, which factor across $\mathbf{E}(\mathbf{T})_{z_R}$ iff $A = \bar{f}^*(R)$ is local. Much more generally, let T be *any* geometric theory that is constructed from the finitely presented algebraic theory T_0 by enriching the language of T_0 to the full geometric language and adding any *finite* number of geometric axioms. Then the process described previously gives the classifying topos $\mathbf{E}(T) \rightarrowtail \mathbf{E}(\mathbf{T}_0)$ for models of T in topoi defined over **E**. In fact, the construction of $\mathbf{E}(\mathbf{T}_0)$ can also be considerably simplified, since neither the products nor the equations are necessary any longer. The only reason this method does not work for a general, finitely presented, geometric theory T is that, as yet, we do not have a sufficiently internal definition of "the Horn part of T." In the case of Grothendieck topoi over **S**, however, this has been elegantly worked out—in the completely general case— by Michel and Marie Françoise Coste (seminar talk, spring 1975, University of Paris VII). All of this goes to show that, when working with topoi, we should not allow the rigidity of the category of sets to dominate our thinking. Constructions such as the Zariski topology associated to a ring are meaningless in **S**, but become very meaningful indeed in topoi with more freedom of movement. Further applications of this method—to the existence of the spectrum of one geometric theory versus another—will be treated in a forthcoming paper with Julian C. Cole.

REFERENCES

[1] M. Coste, Logique du 1er ordre dans les topos élémentaires, *Seminaire Benabou* 1973–1974 (exposé multigraphé).

[2] R. Diaconescu, Change of base for some toposes, *J. Pure Appl. Algebra* (to appear).

[3] M. Hakim, Topos annelés et schémas relatifs, *Ergebnisse Math.* **64** Springer-Verlag, Berlin, 1972.

[4] P. T. Johnstone, Internal category theory, Thesis, Cambridge University 1974.

[5] F. W. Lawvere, Quantifiers and sheaves, *Actes, Congrès Intern. Math.* (1970), tome I, pp. 329–334.

[6] F. W. Lawvere, Continuously variable sets: algebraic geometry = geometric logic, *Proc. Logic Colloquium, Bristol* 1973.

[7] B. Lesaffre, Structures algébriques dans les topos élémentaires, Thèse de 3e cycle, Université Paris VII, 1974.

[8] M. Tierney, Axiomatic sheaf theory, *Proc. CIME Conf. Category Theory and Commutative Algebra, Varenna 1971, Edizione Cremonese 1973*, pp. 249–326.

AMS 18F20

DEPARTMENT OF MATHEMATICS
RUTGERS UNIVERSITY
NEW BRUNSWICK, NEW JERSEY

and

UNIVERSITÉ PARIS VII

Forcing Topologies and Classifying Topoi

MYLES TIERNEY

This paper is a sequel, or companion to [5], which also appears in this volume. There I introduced the concept of "forcing topology" and used it to define the Zariski topology associated to a commutative ring A in an arbitrary topos **E**. I knew there was a strong connection between these forcing topologies and the existence of classifying topoi, but due to time pressure—[5] was submitted May 9th, 1975—I was not quite able to make this connection explicit. A short time afterwards, I succeeded in constructing the classifying topos for a finitary geometric theory over an arbitrary base topos with a natural numbers object. I first talked about this, and the absurdly simple construction of the spectrum of a ringed topos described below, in Benabou's seminar on May 24th. Benabou, in the course of three lectures beginning a week later, then added several very pertinent observations, which placed the construction defined here in a much more general context. I will make only a few bare remarks concerning these, since I am sure Benabou will wish to develop this material himself. He also noticed, independently, the existence of classifying topoi for finite limits and **S**—indexed colimits, modules, categories, functors, distributors, flat distributors, and the exponential topos. Rather than artificially rewriting [5], it seemed better to treat classifying topoi in a separate paper. In order to make the treatment self-contained, we discuss forcing topologies in Section 1, although this duplicates some material of [5].

1. Forcing Topologies

Let **E** be an arbitrary topos. If $I \rightarrowtail \Omega$, then, as is well-known by now (see [2] for the original treatment), there is a smallest $J \rightarrowtail \Omega$, such that the characteristic map of J is a topology and $I \subseteq J$. In fact, J is the internal inf of all such subobjects of Ω; the associated topology is called the *topology generated by I*. Now if $X' \overset{m}{\rightarrowtail} X$ is an arbitrary monomorphism of **E** with characteristic map $\varphi\colon X \to \Omega$, let j_m be the topology generated by the image $I_m \rightarrowtail \Omega$ of φ, and denote by $J_m \rightarrowtail \Omega$ the object of j_m-dense subobjects of 1. j_m is called the *forcing topology* for m, because it has the following universal property.

If $f\colon \mathbf{E}_0 \to \mathbf{E}$ is a geometric morphism, and j_f is the topology in **E** whose category of sheaves is the image of f, write J_f for its object of j_f-dense subobjects of 1. Then f factors across $\mathbf{E}_{j_m} \rightarrowtail \mathbf{E}$ iff

$$\begin{array}{ccc} \mathbf{E}_{j_f} & \dashrightarrow & \mathbf{E}_{j_m} \\ & \searrow \quad \swarrow & \\ & \mathbf{E} & \end{array}$$

iff $J_m \subseteq J_f$, iff $I_m \subseteq J_f$, iff $f^*(m)$ is an isomorphism. Note this shows that

$$\begin{array}{ccc} \mathbf{E}_{0 j_{f^*(m)}} & \longrightarrow & \mathbf{E}_{j_m} \\ \downarrow & & \downarrow \\ \mathbf{E}_0 & \xrightarrow[f]{} & \mathbf{E} \end{array}$$

is a pullback. Also, any topology arises in this way by taking m to be $1 \overset{\text{true}}{\rightarrowtail} J$.

If $X' \overset{m}{\rightarrowtail} X$ and $Y' \overset{n}{\rightarrowtail} Y$ is an arbitrary pair of monomorphisms, let $j_{m\,\&\,n}$ be the topology generated by $I_m \vee I_n \rightarrowtail \Omega$. Then $\mathbf{E}_{j_{m\,\&\,n}} \rightarrowtail \mathbf{E}$ is $\mathbf{E}_{j_m} \wedge \mathbf{E}_{j_n} \rightarrowtail \mathbf{E}$, and a geometric morphism f factors across this iff f^* inverts both m and n.

Let $k\colon X \to Y$ be an arbitrary morphism of **E**. Then k has the canonical factorization

$$\begin{array}{ccccccc} X & \overset{\delta}{\rightarrowtail} & X \underset{Y}{\times} X & \rightrightarrows & X & \overset{k}{\longrightarrow} & Y \\ & & & & & \searrow \quad \nearrow_{\lambda} & \\ & & & & & Q & \end{array}$$

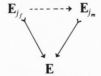

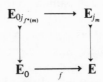

where δ is the diagonal. k is monic iff δ is an isomorphism, epic iff λ is an isomorphism, and an isomorphism iff both δ and λ are isomorphisms. Thus there are three topologies associated to k: j_δ, which forces k to be monic; j_λ, which forces it to be epic; and $j_{\delta\&\lambda}$, which forces it to be an isomorphism.

In the sequel we will use forcing topologies to construct models of a theory in the following way. Let T be a finitary geometric theory. That is, T has a finite number of sorts i, a finite number of function symbols $f: i_1, \ldots, i_n \to i$, and a finite number of atomic formulas $r(j_1, \ldots, j_m)$. The rest of the formulas of T are constructed using the logical connectives true, $\wedge$, $=$, false, and $\vee$, together with the existential quantifier $\exists$—i.e. we do *not* use $\neg$, $\Rightarrow$, or $\forall$. The axioms of T consist of a finite number of sequents $\alpha \vdash \beta^V$, where α and β are formulas of T, and V is a finite set of variables. Rules for manipulating these can be found in [1].

To T we associate a certain finite diagram $\mathbb{D}_T$; for each sort i we take an object d_i. For each function symbol f we take a new object d_f together with one morphism $d_f \to d_i$ and n morphisms $p_l: d_f \to d_{i_l}$, $l = 1, \ldots, n$. For each atomic formula r we take a new object d_r together with m morphisms $q_k: d_r \to d_{i_k}$, $k = 1, \ldots, m$.

Suppose $M: \mathbb{D}_T \to \mathbf{E}$ is a diagram of type $\mathbb{D}_T$ in $\mathbf{E}$. Then in $\mathbf{E}$ we have a finite number of morphisms

$$M_f \to M_{i_1} \times \cdots \times M_{i_n} \qquad \text{and} \qquad M_r \to M_{j_1} \times \cdots \times M_{j_m}.$$

Let j_S be the topology that forces the first group to be isomorphisms, and the second to be monomorphisms. In $\mathbf{E}_{j_S}$, M becomes a structure for T, and if $\mathbf{E}_0 \xrightarrow{f} \mathbf{E}$ is a geometric morphism, f factors across $\mathbf{E}_{j_S} \rightarrowtail \mathbf{E}$ iff f^*M is a structure for T in $\mathbf{E}_0$. In $\mathbf{E}_{j_S}$, for each axiom $\alpha \vdash \beta^V$ we have an interpretation

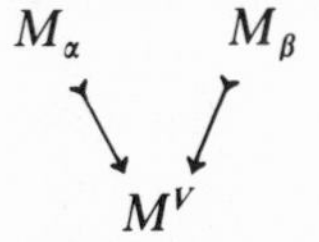

where M^V means $M_{i_1} \times \cdots \times M_{i_k}$ if the types of the variables of V are $i_1, \ldots, i_k$. Now $\alpha \vdash \beta^V$ is valid iff $M_\alpha \subseteq M_\beta$, iff in

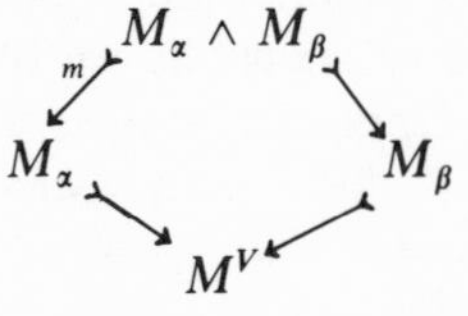

m is an isomorphism. If j_T is the topology that forces all of these m's to be isomorphisms, then M becomes a model for T in $(\mathbf{E}_{j_S})_{j_T}$. Thus, taking j_M to be the topology in $\mathbf{E}$ generated by the composite

$$(\mathbf{E}_{j_S})_{j_T} \rightarrowtail \mathbf{E}_{j_S} \rightarrowtail \mathbf{E}$$

we have forced M to be a model of T in $\mathbf{E}_{j_M} \simeq (\mathbf{E}_{j_S})_{j_T}$, and a geometric morphism $f: \mathbf{E}_0 \rightarrowtail \mathbf{E}$ factors across $\mathbf{E}_{j_M} \rightarrowtail \mathbf{E}$ iff f^*M is a model of T in $\mathbf{E}_0$. Again, this shows that

$$\begin{array}{ccc} \mathbf{E}_{0j_{f^*M}} & \longrightarrow & \mathbf{E}_{j_M} \\ \big\downarrow & & \big\downarrow \\ \mathbf{E}_0 & \xrightarrow{\quad f \quad} & \mathbf{E} \end{array}$$

is a pullback.

There are two remarks to make concerning this process of forcing. One is important, while the other is merely stylistic.

For the latter, no importance should be attached to the fact that we have separated this process into two steps—this is only for reasons of clarity. That is, often one is dealing directly with a structure for T, so the first step is unnecessary. For example, if T is the theory of local rings [5], and A is a commutative ring in $\mathbf{E}$, then A is already a structure for T, and the Zariski topology z_A is the topology that forces the two extra axioms

$$\text{true} \vdash \exists a'(a'a = 1) \vee \exists a''(a''(1 - a) = 1)^{\{a\}}$$

and

$$(0 = 1) \vdash \text{false}.$$

For the former, now that we know how to force diagrams to become models, it is clear that to construct the classifying topos, we will simply force the universal diagram to be a model. Thus, we need to know how to classify finite diagrams. This, as we will see, is very easy.

I should like to stress, however, that we have concentrated on diagrams of the form $\mathbb{D}_T$ only because we are interested here in models of the theory T. In fact, given any diagram $\mathbb{D}$ in a topos $\mathbf{E}$ we can force any appropriate finite configuration (or even not necessarily finite if properly indexed over a base topos) in $\mathbb{D}$ to become a limit or colimit. Of course, if done inconsistently this will yield the empty topos; but no matter, it can be done. When we do it in the universal example, we obtain the classifying topos for this notion—*anything* preserved by inverse images can be classified. As remarked in the introduction, this has also been observed by Benabou.

2. The Classifying Topos

To classify finite diagrams, we must, in particular, be able to classify objects, so we place ourselves in the context of a base topos $\mathbf{S}$ with an object of natural numbers N. Following Wraith [6], one can then form

$\mathscr{S}_{\text{fin}}$, the full internal subcategory of finite objects of $\mathbf{S}$. I recall that if $[n] = N \times N \xrightarrow{+} N \xrightarrow{s} N$ is the generic finite object in $\mathbf{S}/N$, then $\mathscr{S}_{\text{fin}}$ is given by taking

$$\langle d_0, d_1 \rangle : (\mathscr{S}_{\text{fin}})_1 \to (\mathscr{S}_{\text{fin}})_0 \times (\mathscr{S}_{\text{fin}})_0$$

to be

$$(\pi_2{}^*[n])^{\pi_1{}^*[n]} \to N \times N$$

in $\mathbf{S}/(N \times N)$—composition is given by evaluation. $\mathscr{S}_{\text{fin}}$ is preserved by the inverse image of a geometric morphism, and $\mathbf{S}(\mathscr{S}_{\text{fin}}) \xrightarrow{\Gamma} \mathbf{S}$ classifies objects. That is, if $\Phi\colon \mathbf{E} \to \mathbf{S}$ is a topos over $\mathbf{S}$, then geometric morphisms

$$\begin{array}{ccc} \mathbf{E} & \xrightarrow{f} & \mathbf{S}(\mathscr{S}_{\text{fin}}) \\ & {\scriptstyle\Phi}\searrow \quad \swarrow{\scriptstyle\Gamma} & \\ & \mathbf{S} & \end{array}$$

are in 1–1 correspondence with objects of $\mathbf{E}$, and the correspondence is given by evaluating f^* on the universal object, which is the "inclusion" of $\mathscr{S}_{\text{fin}}$ in $\mathbf{S}$.

Now, if $\mathbf{B}$ is a category with finite limits, $\mathbf{C} \in \text{Cat}(\mathbf{B})$, and $\mathbb{D}$ is a finite diagram, there is a general construction, due to Benabou, which yields a new category $\mathbf{C}^{\mathbb{D}}$ in $\mathbf{B}$. Let us outline it in the case $\mathbf{B} = \mathbf{S}$ and $\mathbf{C} = \mathscr{S}_{\text{fin}}$. First the case $\mathbb{D} = \mathbb{2}$. We take $(\mathscr{S}^{\mathbb{2}}_{\text{fin}})_0$ to be $(\mathscr{S}_{\text{fin}})_1$, and $(\mathscr{S}^{\mathbb{2}}_{\text{fin}})_1$ to be the object of commutative squares of $\mathscr{S}_{\text{fin}}$. Since an arbitrary finite $\mathbb{D}$ is a finite colimit, in the category of diagrams, of pieces involving only $\mathbb{1}$ and $\mathbb{2}$, we take $\mathscr{S}^{\mathbb{D}}_{\text{fin}}$ to be the *limit* of the corresponding diagram whose pieces are $\mathscr{S}^{\mathbb{1}}_{\text{fin}} = \mathscr{S}_{\text{fin}}$ and $\mathscr{S}^{\mathbb{2}}_{\text{fin}}$.

$\mathbf{S}(\mathscr{S}^{\mathbb{D}}_{\text{fin}}) \to \mathbf{S}$ is the diagram classifier. Why is this? Well, for $\mathbf{S}$ the category of sets, it is obvious, because $(\mathscr{S}^{\mathbb{D}}_{\text{fin}})^{\text{op}}$ is clearly the finite limit completion of $\mathbb{D}$. That is, we have a universal diagram

$$\mathbb{D} \to (\mathscr{S}^{\mathbb{D}}_{\text{fin}})^{\text{op}},$$

which is such that if $\Phi\colon \mathbf{E} \to \mathbf{S}$ is defined over $\mathbf{S}$, and $D\colon \mathbb{D} \to \mathbf{E}$ is a diagram in $\mathbf{E}$, then there is a unique left exact functor $\overline{D}$ such that

$$\begin{array}{ccc} \mathbb{D} & \longrightarrow & (\mathscr{S}^{\mathbb{D}}_{\text{fin}})^{\text{op}} \\ & {\scriptstyle D}\searrow \quad \swarrow{\scriptstyle\overline{D}} & \\ & \mathbf{E} & \end{array}$$

commutes. But from Diaconescu's theorem [2], we know that

$$\text{Lex}((\mathscr{S}^{\mathbb{D}}_{\text{fin}})^{\text{op}}, \mathbf{E}) \simeq \text{Top}_{\mathbf{S}}(\mathbf{E}, \mathbf{S}(\mathscr{S}^{\mathbb{D}}_{\text{fin}})).$$

Now the last part remains exactly as is over an arbitrary $\mathbf{S}$, and we can internalize the first part. Another way of approaching this is to notice that since the passage $\mathbb{D} \rightsquigarrow (\mathscr{S}_{\mathrm{fin}}^{\mathbb{D}})^{\mathrm{op}}$ is going to be a left adjoint, and $\mathbb{D}$ is a colimit of $\mathbb{1}$'s and $\mathbb{2}$'s, $(\mathscr{S}_{\mathrm{fin}}^{\mathbb{D}})^{\mathrm{op}}$ is also a *colimit* in Lex—the category of finitely complete categories and left exact functors. By Diaconescu's theorem above this must go into a *limit* in $\mathrm{Top}_{\mathbf{S}}$ of pieces of the form $\mathbf{S}(\mathscr{S}_{\mathrm{fin}}^{\mathbb{2}}) \to \mathbf{S}$ and $\mathbf{S}(\mathscr{S}_{\mathrm{fin}}) \to \mathbf{S}$, so we need only prove that $\mathbf{S}(\mathscr{S}_{\mathrm{fin}}^{\mathbb{2}}) \to \mathbf{S}$ is the morphism classifier, which is easy.

Benabou has pointed out that the pair

$$\mathbf{S}(\mathscr{S}_{\mathrm{fin}}^{\mathbb{2}}) \underset{d_1}{\overset{d_0}{\rightrightarrows}} \mathbf{S}(\mathscr{S}_{\mathrm{fin}}),$$

which classifies the domain and codomain of the universal morphism, yields a category $\mathbf{C}$ in $\mathrm{Top}_{\mathbf{S}}$. This is clear from the above analysis, since $\mathbb{1} \rightrightarrows \mathbb{2}$ is a *cocategory*, as is well-known. But now this category represents $\mathrm{Top}_{\mathbf{S}}$, and the really important, and very pretty, observation of Benabou's is that in this way the construction of the diagram classifier, once one has the morphism classifier, becomes the construction of $\mathbf{C}^{\mathbb{D}}$ in $\mathrm{Top}_{\mathbf{S}}$. Moreover, the same holds for the general classifying topos—it is essentially a construction involving a geometric category in a category with finite limits—since the representing category $\mathbf{C}$ exists, all the forcing topologies take place there, and their categories of sheaves can be described as certain equalizers in $\mathrm{Top}_{\mathbf{S}}$. In this way, by pulling back, all forcing topologies are equalizers. Note that this viewpoint requires a base topos with a natural numbers object, however, while forcing topologies are quite independent of this.

In any case, let me reiterate explicitly that we are now finished with the construction of the classifying topos. For if T is a finitary geometric theory, then $\mathbf{S}(\mathscr{S}_{\mathrm{fin}}^{\mathbb{D}_T}) \to \mathbf{S}$ classifies diagrams of type $\mathbb{D}_T$ in topoi $\mathbf{E} \to \mathbf{S}$. If we force the universal diagram in $\mathbf{S}(\mathscr{S}_{\mathrm{fin}}^{\mathbb{D}_T})$ to become a model of T, yielding $\mathbf{S}_T \rightarrowtail \mathbf{S}(\mathscr{S}_{\mathrm{fin}}^{\mathbb{D}_T})$, then $\mathbf{S}_T \rightarrowtail \mathbf{S}$ classifies diagrams of type $\mathbb{D}_T$, which are models of T; i.e., it classifies models of T. If $f\colon \mathbf{E} \to \mathbf{S}$ is a geometric morphism, then $f^*(\mathscr{S}_{\mathrm{fin}}^{\mathbb{D}_T}) \simeq \mathscr{E}_{\mathrm{fin}}^{\mathbb{D}_T}$, so by the results of Section 1 the classifying topos is stable under change of base; that is,

$$\begin{array}{ccc} \mathbf{E}_T & \longrightarrow & \mathbf{S}_T \\ \big\downarrow & & \big\downarrow \\ \mathbf{E} & \xrightarrow{\ f\ } & \mathbf{S} \end{array}$$

is a pullback.

With regard to previous, and future, work in this area, it is clear that this construction is simpler and more general than those of Johnstone [3] or Wraith [6], who treat only the finite algebraic case. In particular, we do not need the existence of free structures. Also, the role of recursion and the natural numbers object is clearly pinpointed—they are necessary *only* to construct the object classifier.

If $\mathbf{S}$ is the category of sets, a slight modification of the above procedure yields the classifying topos of Makkai–Reyes [4] for an arbitrary geometric theory T. Namely, since we have arbitrary small colimits, we have no trouble in handling infinite disjunctions or in forcing arbitrary diagrams of type $\mathbb{D}_T$ (same definition) to become models. To classify diagrams, we consider $\mathbb{D} \to (\mathbf{S}^{\mathbb{D}})^{\mathrm{op}}$ and take $(\mathbf{S}^{\mathbb{D}})^{\mathrm{op}}_{\mathrm{fin}}$, which is the closure of $\mathbb{D}$ under finite limits. Better still, use the morphism classifier and the existence of set indexed limits of Grothendieck topoi. These are not hard to construct using what follows. Obviously, the same procedure is going to work over an arbitrary base, as soon as we make precise what an internal geometric theory is. Here Benabou also has made a good start. Even now, however, we can classify a good deal more than just models of a finitary geometric theory.

As remarked earlier, we can classify any type of finite limit or $\mathbf{S}$-indexed colimit. Or, take the constructions of [5]. There if A is a commutative ring in $\mathbf{S}$, I defined an order on A such that $\mathbf{S}(\mathbf{A}^{\mathrm{op}}) \to \mathbf{S}$ classified saturated multiplicative subobjects of A. With the present technique such a construction is trivial. Namely, let d_1 classify the codomain of the universal morphism: $d_1\colon \mathbf{S}(\mathscr{S}^{2}_{\mathrm{fin}}) \to \mathbf{S}(\mathscr{S}_{\mathrm{fin}})$. Let $\tilde{A}\colon \mathbf{S} \to \mathbf{S}(\mathscr{S}_{\mathrm{fin}})$ classify the object A. In the pullback

$$\begin{array}{ccc} \mathbf{P} & \longrightarrow & \mathbf{S}(\mathscr{S}^{2}_{\mathrm{fin}}) \\ {\scriptstyle\Gamma}\downarrow & & \downarrow{\scriptstyle d_1} \\ \mathbf{S} & \xrightarrow{\ \tilde{A}\ } & \mathbf{S}(\mathscr{S}_{\mathrm{fin}}) \end{array}$$

$\mathbf{P}$ classifies morphisms with codomain A—i.e., if $\Phi\colon \mathbf{E} \to \mathbf{S}$, the maps

$$\begin{array}{ccc} \mathbf{E} & \xrightarrow{\ f\ } & \mathbf{P} \\ & {\scriptstyle\Phi}\searrow \quad \swarrow{\scriptstyle\Gamma} & \\ & \mathbf{S} & \end{array}$$

are in 1–1 correspondence with morphisms $S \to \Phi^*A$. Now, in $\mathbf{P}$ we have the universal morphism $S \to \Gamma^*A$, which we first force to be monic. Then, since the axioms for a saturated multiplicative subobject [5] are geometric, just force these in the universal monomorphism $S \rightarrowtail \Gamma^*A$. This yields $\mathbf{S}(\mathbf{A}^{\mathrm{op}}) \to \mathbf{S}$, and the spectrum is obtained as sheaves for the Zariski topology $z_{(\Gamma^*A)_S}$. Even quicker, just notice that the four axioms for a prime [5] are also geometric, and force *these* in the universal monomorphism. This gives $\mathrm{Spec}(\mathbf{S}, A)$ immediately. Many other examples of the spectrum process follow similarly, and as indicated in [5], will be treated in a separate paper with Julian Cole. Along the above lines, take the theory of A-modules (A not necessarily commutative) or G-objects for G a group in $\mathbf{S}$. The theory of rings is clearly algebraic, as is the theory of modules (with variable ring) so we have classifying topoi Modules $\to \mathbf{S}$ and Rings $\to \mathbf{S}$ together with the morphism Modules $\to$ Rings that classifies the universal ring for the universal module. A-modules are classified by the pullback

$$\begin{array}{ccc} A\text{-Modules} & \longrightarrow & \text{Modules} \\ \downarrow & & \downarrow \\ \mathbf{S} & \xrightarrow{\tilde{A}} & \text{Rings} \end{array}$$

and similarly for G-objects.

In the same vein, we can classify categories, functors, natural transformations, or distributors—each with or without fixed domain or codomain. Discrete fibrations (with fixed base) are functors satisfying a condition described by certain finite limits, so we can classify these. The definition of filtered categories stipulates that certain maps defined by finite limits be epic, so we can classify these. Together we can classify flat functors—Diaconescu's theorem shows this is just another construction of the functor category. But we can go further and classify flat distributors, even continuous flat distributors between fixed sites. This last gives the exponential topos for a pair of bounded topoi over $\mathbf{S}$. When $\mathbf{S}$ = sets, this idea is due to Joyal, who discussed it in a talk at the open house in Sussex in 1974. The continuity condition has been made precise by Cole. In any case, since we are over an *arbitrary* base topos here, we have the construction of the $\prod$ operation for bounded topoi as the usual equalizer. An easy argument from this yields sets-indexed limits for Grothendieck topoi, which is a remark we used earlier.

Note added in proof (*September 1975*)

I first heard about Joyal's result on the exponential topos from Julian Cole in May 1975—I was not present at the Open House myself. Later, in July, I spoke to Joyal in Amiens, and he told me that in this talk at Sussex he had also outlined a general construction of classifying topoi, over the category of sets, that has points in common with this one. For example, though I have not seen any details, I believe his use of forcing topologies is essentially the same as mine. Also, he remarked that some finiteness condition, such as coherence, on the exponent of the exponential topos is probably necessary to express the continuity condition. This does not affect the construction of sets-indexed limits of Grothendieck topoi. In addition,

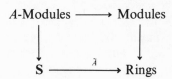

I would like to add that the remark that Johnstone's thesis [3] is concerned only with classifying topoi for finite algebraic theories is not quite accurate, as pointed out to me by Johnstone. He does, in fact, consider the case of a classifying topos for the theory of categories.

REFERENCES

[1] M. Coste, Logique du 1er ordre dans les topos élémentaires, *Seminaire Benabou* 1973–74 (exposé multigraphé).

[2] R. Diaconescu, Change of base for some toposes, *J. Pure Appl. Algebra* (to appear).

[3] P. T. Johnstone, Internal category theory, Thesis, Cambridge University, 1974.

[4] M. Makkai and G. Reyes, Model-theoretic methods in the theory of topoi, *Tagungsbericht Oberwolfach* 1974.

[5] M. Tierney, On the spectrum of a ringed topos (this volume).

[6] G. C. Wraith, Algebraic theories in topoi, preprint, University of Sussex, 1974.

AMS 18F20

DEPARTMENT OF MATHEMATICS
RUTGERS UNIVERSITY
NEW BRUNSWICK, NEW JERSEY

and

UNIVERSITÉ PARIS VII

Published Works of Samuel Eilenberg

Books

1. [with N. E. Steenrod] "Foundations of Algebraic Topology." Princeton Univ. Press, Princeton, New Jersey, 1952.
2. [with H. Cartan] "Homological Algebra." Princeton Univ. Press, Princeton, New Jersey, 1956.
3. [with C. C. Elgot] "Recursiveness." Academic Press, New York, 1970.
4. "Automata, Languages, and Machines," Vol. A. Academic Press, New York, 1974.
5. "Automata, Languages, and Machines," Vol. B. Academic Press, New York, 1976.

Papers

1. Remarques sur les ensembles et les fonctions relativement mésurables, *C. R. Soc. Sci. Varsovie*, Chap. III, **25** (1932), 93–98.
2. Sur les transformations périodiques de la surface de sphère, *Fund. Math.* **22** (1934), 28–41.
3. Sur les transformations continues d'éspaces metriques compacts, *Fund. Math.* **22** (1934), 292–296.
4. Sur les décompositions des continus en ensembles connexes, *Fund. Math.* **22** (1934), 297–302.
5. Sur quelques propriétés des transformations localement homéomorphes, *Fund. Math.* **24** (1935), 35–42.
6. Sur le plongement des éspaces dans les continus acycliques, *Fund. Math.* **24** (1935), 65–71.
7. Deux théoremes sur l'homologie dans les éspaces compacts, *Fund. Math.* **24** (1935), 151–155.

8. Remarque sur un théoreme de M. W. Hurewicz, *Fund. Math.* **24** (1935), 156–159.
9. Sur les transformations d'éspaces metriques en circonférence, *Fund. Math.* **24** (1935), 160–176.
10. Sur l'invariance par rapport aux petites transformations, *C. R. Acad. Sci. Paris* **200** (1935), 1003–1005.
11. [with S. Saks] Sur la dérivation des fonctions dans des ensembles denombrables, *Fund. Math.* **25** (1935), 264–266.
12. Sur quelques proprietes topologiques de la surface de sphère, *Fund. Math.* **25** (1935), 267–272.
13. Ozastosowaniach topologicznych odwzcicwan no okrag kola, *Wiadom. Mat.* **41** (1935), 1–32.
14. Transformations continues en circonférence et la topologie du plan, *Fund. Math.* **26** (1936), 61–112.
15. Sur le théoreme de décomposition de la théorie de la dimension, *Fund. Math.* **26** (1936), 146–149.
16. [with K. Borsuk] Uber stetige Abbildungen der Teilmengen euklidischer Raume auf die Kreislinie, *Fund. Math.* **26** (1936), 207–223.
17. Bemerkungen zur Pontrjagin'schen Verallgemeinerung des Alexander'schen Dualitatssatzes, *Fund. Math.* **26** (1936), 224–228.
18. Un théoreme de dualités, *Fund. Math.* **26** (1936), 280–282.
19. Sur les éspaces multicohérents I, *Fund. Math.* **27** (1936), 153–190.
20. Sur un théoreme topologique de M. L. Schnirelmann, *Rec. Math. Moscou*, **1** (1936), 557–560.
21. Uber ein Problem von H. Hopf, *Fund. Math.* **28** (1937), 58–60.
22. Sur les groupes compacts d'homéomorphies, *Fund. Math.* **28** (1937), 75–80.
23. Sur les courbes sans noeuds, *Fund. Math.* **28** (1937), 233–242.
24. Sur l'enlacement faible, *C. R. Acad. Sci. Paris* **204** (1937), 1226–1227.
25. Sur les éspaces multicohérents II, *Fund. Math.* **29** (1937), 101–122.
26. Sur les ensembles plans localement connexes, *Fund. Math.* **29** (1937), 159–160.
27. Un théoreme sur l'homotopie, *Ann. of Math.* **38** (1937), 656–661.
28. Sur la multicohérence des surfaces closes, *C. R. Soc. Sci. Varsovie*, Chap. III, **30** (1937), 109–111.
29. Sur les transformations à petites tranches, *Fund. Math.* **30** (1938), 92–95.
30. [with E. Otto] Quelques propriétés characteristiques de la dimension, *Fund. Math.* **31** (1938), 149–153.
31. Sur le prolongement des transformations en surfaces sphériques, *Fund. Math.* **31** (1938), 179–200.
32. On φ-measures, *Ann. Soc. Pol. Math.* **17** (1938), 251–252.
33. On continua of finite length 1, *Ann. Soc. Pol. Math.* **17** (1938), 253–254.
34. Cohomologies et transformations continues, *C. R. Acad. Sci. Paris* **208** (1939), 68–69.
35. Généralisation du théoreme de M. H. Hopf sur les classes des transformations en surfaces sphériques, *Compositio Math.* **6** (1939), 428–433.
36. On the relation between the fundamental group of a space and the higher homotopy groups, *Fund. Math.* **32** (1939), 167–175.
37. [with C. Kuratowski] Théoremes d'addition concernant le groupe des transformations en circonference, *Fund. Math.* **39** (1939), 193–200.
38. Cohomology and continuous mappings, *Ann. of Math.* **41** (1940), 231–251.
39. On continuous mappings of manifolds into spheres, *Ann. of Math.* **41** (1940), 662–673.
40. On a theorem of P. A. Smith concerning fixed points for periodic transformations, *Duke Math. J.* **6** (1940), 428–437.

41. On homotopy groups, *Proc. Nat. Acad. Sci.* **26** (1940), 563–565.
42. Ordered topological spaces, *Amer. J. Math.* **63** (1941), 39–45.
43. An invariance theorem for subsets of S^n, *Bull. Amer. Math. Soc.* **47** (1941), 73–75.
44. Continuous mappings of infinite polyhedra, *Ann. of Math.* **42** (1941), 459–468.
45. On spherical cycles, *Bull. Amer. Math. Soc.* **47** (1941), 432–434.
46. Extensions and classification of continuous mappings, "Lectures in Topology." Univ. of Michigan Press, Ann Arbor, (1941), 57–99.
47. [with S. Mac Lane] Infinite cycles and homologies, *Proc. Nat. Acad. Sci.* **27** (1941), 535–539.
48. [with E. W. Miller] Zero-dimensional families of sets, *Bull. Amer. Math. Soc.* **47** (1941), 921–923.
49. Banach space methods in topology, *Ann. of Math.* **43** (1942), 568–579.
50. [with R. L. Wilder] Uniform local connectedness and contractibility, *Amer. J. Math.* **64** (1942), 613–622.
51. [with S. Mac Lane] Group extensions and homology, *Ann. of Math.* **43** (1942), 757–831.
51a. [with S. Mac Lane] Appendix to S. Lefschetz's "Algebraic Topology," pp. 344–349. *Amer. Math. Soc.*, Providence, Rhode Island, 1942.
52. [with S. Mac Lane] Natural isomorphisms in group theory, *Proc. Nat. Acad. Sci.* **28** (1942), 537–543.
53. [with O. G. Harrold] Continua of finite linear measure, *Amer. J. Math.* **65** (1943), 137–146.
54. [with S. Mac Lane] Relations between homology and homotopy groups, *Proc. Nat. Acad. Sci.* **29** (1943), 155–158.
55. [with I. Niven] The "fundamental theorem of algebra" for quaternions, *Bull. Amer. Math. Soc.* **50** (1944), 246–248.
56. Continua of finite linear measure II, *Amer. J. Math.* **66** (1944), 425–427.
57. Singular homology theory, *Ann. of Math.* **45** (1944), 407–447.
58. [with N. E. Steenrod] Axiomatic approach to homology theory, *Proc. Nat. Acad. Sci.* **31** (1945), 117–120.
59. [with S. Mac Lane] General theory of natural equivalences, *Trans. Amer. Math. Soc.* **58** (1945), 231–294.
60. [with S. Mac Lane] Relations between homology and homotopy groups of spaces, *Ann. of Math.* **46** (1945), 480–509.
61. [with D. Montgomery] Fixed point theorem for multi-valued transformations, *Amer. J. Math.* **68** (1946), 480–509.
62. [with S. Mac Lane] Determination of the second homology and cohomology groups of a space by means of homotopy invariants, *Proc. Nat. Acad. Sci.* **32** (1946), 277–280.
63. [with S. Mac Lane] Cohomology theory in abstract groups I, *Ann. of Math.* **48** (1947), 51–78.
64. [with S. Mac Lane] Cohomology theory in abstract groups II, Group extensions with a non-abelian kernel, *Ann. of Math.* **48** (1947), 326–341.
65. Homology of spaces with operators I, *Trans. Amer. Math. Soc.* **61** (1947), 378–417.
66. [with S. Mac Lane] Algebraic cohomology groups and loops, *Duke Math. J.* **14** (1947), 435–463.
67. Singular homology in differentiable manifolds, *Ann. of Math.* **48** (1947), 670–681.
68. On a linkage theorem by L. Cesari, *Bull. Amer. Math. Soc.* **53** (1947), 1192–1195.
69. [with C. Chevalley] Cohomology theory of Lie groups and Lie algebras, *Trans. Amer. Math. Soc.* **63** (1948), 85–124.
70. Relations between cohomology groups in a complex, *Comment. Math. Helv.* **21** (1948), 302–320.

71. [with S. Mac Lane] Cohomology and Galois theory I, Normality of algebras and Teichmuller's cocycle, *Trans. Amer. Math. Soc.* **64** (1948), 1–20.
72. Extensions of general algebras, *Ann. Soc. Polonaise Math.* **21** (1948), 125–134.
73. [with S. Mac Lane] Homology of spaces with operators II, *Trans. Amer. Math. Soc.* **65** (1949), 49–99.
74. Topological methods in abstract algebra, Cohomology theory of groups, *Bull. Amer. Math. Soc.* **55** (1949), 3–37.
75. On the problems of topology, *Ann. of Math.* **50** (1949), 247–260.
76. [with J. A. Zilber] Semi-simplicial complexes and singular homology, *Ann. of Math.* **51** (1950), 499–513.
77. [with S. Mac Lane] Relations between homology and homotopy groups of spaces II, *Ann. of Math.* **51** (1950), 514–533.
78. [with S. Mac Lane] Cohomology theory of abelian groups and homotopy theory I, *Proc. Nat. Acad. Sci.* **36** (1950), 443–447.
79. [with S. Mac Lane] Cohomology theory of abelian groups and homotopy theory II, *Proc. Nat. Acad. Sci.* **36** (1950), 657–663.
80. [with S. Mac Lane] Cohomology theory of abelian groups and homotopy theory III, *Proc. Nat. Acad. Sci.* **37** (1951), 307–310.
81. [with S. Mac Lane] Homology theories for multiplicative systems, *Trans. Amer. Math. Soc.* **71** (1951), 294–330.
82. [with S. Mac Lane] Cohomology theory of abelian groups and homotopy theory IV, *Proc. Nat. Acad. Sci.* **38** (1952), 325–329.
82a. Homotopy groups and algebraic homology theories, *Proc. Int. Congress 1950*, 349–353.
83. [with S. Mac Lane] Acyclic models, *Amer. J. Math.* **75** (1953), 189–199.
84. [with J. A. Zilber] On products of complexes, *Amer. J. Math.* **75** (1953), 200–204.
85. [with S. Mac Lane] On the groups $H(\Pi, n)$, *Ann. of Math.* **58** (1953), 55–106.
86. [with S. Mac Lane] On the groups $H(\Pi, n)$, II, *Ann. of Math.* **60** (1954), 49–139.
87. [with S. Mac Lane] On the groups $H(\Pi, n)$, III, *Ann. of Math.* **60** (1954), 513–557.
88. [with S. Mac Lane] On the homology theory of abelian groups, *Canad. J. Math.* **7** (1955), 45–53.
89. Algebras of cohomologically finite dimension, *Comment. Math. Helv.* **28** (1954), 310–319.
90. [with H. Ikeda and T. Nakayama] On the dimension of modules and algebras I, *Nagoya Math. J.* **8** (1955), 49–57.
91. [with T. Nakayama] On the dimension of modules and algebras II, Frobenius algebras and quasi-Frobenius rings, *Nagoya Math. J.* **9** (1955), 1–16.
92. [with H. Nagao and T. Nakayama] On the dimension of modules and algebras IV, Dimension of residue rings of hereditary rings, *Nagoya Math. J.* **10** (1956), 87–95.
93. Homological dimension and syzygies, *Ann. of Math.* **64** (1956), 328–336, Errata **65** (1957), 593.
94. [with T. Nakayama] On the dimension of modules and algebras V. Dimension of residue rings, *Nagoya Math. J.* **11** (1957), 9–12.
95. [with T. Ganea] On the Lusternik–Schnirelmann category of abstract groups, *Ann. of Math.* **65** (1957), 517–518.
96. [with A. Rosenberg and D. Zelinsky] On the dimension of modules and algebras VIII, Dimension of tensor products, *Nagoya Math. J.* **12** (1957), 71–93.
97. [with H. Cartan] Foundations of fibre bundles, *Symp. Inter. Topologia Algebraica, Mexico 1958*, 16–23.
97a. Foundations of fiber bundles, *Chicago Notes.*
98. Abstract description of some basic functors, *J. Indiana Math. Soc.* **24** (1960), 231–234.
99. [with J. C. Moore] Limits and spectral sequences, *Topology* **1** (1962), 1–23.

100. [with K. Kuratowski] A remark on duality, *Fund. Math.* **50** (1962), 515–517.
101. [with J. C. Moore] Foundations of relative homological algebra, *Memoirs Amer. Math. Soc.* **55** (1965), 1–39.
102. [with J. C. Moore] Adjoint functors and triples, *Illinois J. Math.* **9** (1965), 381–398.
103. [with J. C. Moore] Homology and fibrations I, *Comment. Math. Helv.* **40** (1956), 199–236.
104. [with J. C. Moore] Homological algebra and fibrations, *Colloq. Topologie, Bruxelles 1966*, 81–90.
105. [with G. M. Kelly] A generalization of the functorial calculus, *J. Algebra* **3** (1966), 366–375.
106. [with G. M. Kelly] Closed categories, *Proc. Conf. Categorical Algebra, La Jolla 1966*, 421–562.
107. [with J. B. Wright] Automata in general algebra, *Information and Control* **II** (1967), 452–470.
108. [with C. C. Elgot] Iteration and recursion, *Proc. Nat. Acad. Sci.* **61** (1968), 378–379.
109. [with M. P. Schutzenberger] Rational sets in commutative monoids, *J. Algebra* **13** (1969), 173–191.
110. [with C. C. Elgot and J. C. Shepherdson] Sets recognized by *n*-tape automata, *J. Algebra* **13** (1969), 447–464.
111. Algebraic aspects of automata theory, *Actes Congrès Inter. Math.* **3** (1970) 265–267.
112. [with E. Dyer] An adjunction theorem for locally equiconnected spaces, *Pacific J. Math.* **41** (1972), 669–685.

A 6
B 7
C 8
D 9
E 0
F 1
G 2
H 3
I 4
J 5